SAS® System for Regression

1986 Edition

SAS Institute Inc.
SAS Circle □ Box 8000
Cary, NC 27512-8000

Regina Luginbuhl edited the *SAS® System for Regression, 1986 Edition.* Copyeditors were **J. Chris Parker** and **David D. Baggett**.

SAS® System for Regression, 1986 Edition

ISBN 1-55544-029-0

1st printing, November 1986
2nd printing, October 1989

Note that text corrections may have been made at each printing.

The Institute is a private company devoted to the support and further development of its software and related services.

Contents

Acknowledgments

We would like to acknowledge several persons at SAS Institute whose efforts have contributed to the completion of this book. First of all, we are grateful to Jim Goodnight who originally encouraged us to write this book. Regina Luginbuhl served as editor. Several persons reviewed the chapters and contributed many useful comments. The reviewers were Margaret Adair, John Boling, John Brocklebank, Eddie Routten, John Sall, Phil Spector, and Maura Stokes.

The work of several persons has influenced our writing. In particular, we acknowledge Professors Walt Harvey of Ohio State University, Ron Hocking of Texas A&M University, Bill Sanders of the University of Tennessee, and Shayle Searle of Cornell University.

Finally, we thank the students at Texas A&M University and the University of Florida whose research projects provided the ideas and data for many of the examples.

About the Authors

Rudolf J. Freund, Ph.D.

Dr. Rudolf J. Freund holds the position of Professor at the Institute of Statistics, Texas A&M University where he has been on the faculty since 1962. He received an M.A. degree in economics from the University of Chicago in 1951 and continued his work for a Ph.D. in statistics at North Carolina State University in 1955. He did early work in statistical computing at Virginia Polytechnic Institute where he taught from 1955 to 1962.

Dr. Freund is the author (with Paul D. Minton) of *Regression Methods* (New York: Marcel Dekker, Inc., 1979) and of "An Introduction to SAS" (available from the author). Currently, Freund teaches a basic statistical methods course and a course on regression for graduate students.

Ramon C. Littell, Ph.D.

Dr. Ramon C. Littell has the rank of Professor of Statistics at the University of Florida in Gainesville. He received his graduate training at Oklahoma State University with an M.S. degree in mathematics in 1966 and a Ph.D. degree in statistics in 1970. Dr. Littell teaches a first-year graduate course in linear models, a course in matrix algebra for statisticians, and a course in data analysis with the SAS System.

Both SAS users since 1972 and former SUGI chairmen, Freund and Littell were closely associated with the Southern Regional Project that funded continued SAS development at North Carolina State University. In addition to their university positions, both authors publish widely in statistical and applied journals, consult in industry, and serve as Consulting Statisticians to their state Agricultural Experiment Stations. They are both Fellows of the American Statistical Association.

Preface

SAS Series in Statistical Applications

The *SAS System for Regression* is one in a series of statistical applications guides developed by SAS Institute. Each book in the SAS Series in Statistical Applications covers a well-defined statistical topic by describing and illustrating relevant SAS procedures.

The following books are currently available:

- *SAS System for Linear Models, 1986 Edition* provides information about features and capabilities of the GLM, ANOVA, REG, MEANS, TTEST, and NESTED procedures. Topics include regression; balanced analysis of variance, including a discussion of the use of multiple-comparison procedures and nested designs; analysis-of-variance models of less-than-full rank; analysis of unbalanced data; analysis of covariance; multivariate linear models; and univariate and multivariate repeated-measures analysis of variance.
- *SAS System for Forecasting Time Series, 1986 Edition* describes how SAS/ETS software can be used to perform univariate and multivariate time series analyses. Early chapters introduce linear regression and autoregression using simple models. Later chapters discuss the ARIMA model and its special applications, state space modeling, spectral analysis, and cross-spectral analysis. The SAS procedures ARIMA, STATESPACE, and SPECTRA are featured, with mention of the simpler procedures FORECAST, AUTOREG, and X11.
- *SAS® System for Elementary Statistical Analysis* teaches you how to perform a variety of data analysis tasks and interpret your results. Written in tutorial style, the guide provides the essential information you need, without overwhelming you with extraneous details. This approach makes the book a ready guide for the business user and an excellent tool for teaching fundamental statistical concepts. Topics include comparing two or more groups, simple regression and basic diagnostics, as well as basic DATA steps.
- *SAS System for Regression, 1986 Edition* is described below.

This series is designed to aid data analysts who use the SAS System and students in statistics courses who want to see more practical applications of the methods discussed in textbooks, lectures, and primary SAS statistical documentation. These manuals are intended to supplement the references you are using now; they will not take the place of a good statistics book and should be used with appropriate SAS user's guides. You will need to be familiar with basic SAS System commands and the DATA step as described in the *SAS Introductory Guide* and the *SAS User's Guide: Basics, Version 5 Edition.*

SAS System for Regression, 1986 Edition

Most statistical analyses are based on linear models, and most analyses of linear models can be performed by three SAS procedures: REG, ANOVA, and GLM. Unlike statistical packages that need different programs for each type of analysis, these procedures provide the power and flexibility for almost all linear model analyses.

To use these procedures properly, you must: (1) understand the statistics you need for the analysis and (2) know how to instruct PROC REG to carry out the computations. The *SAS System for Regression* was written to make it easier for you to apply these procedures to your data analysis problems. In this book, a wide variety of data are used to illustrate the basic kinds of regression models that can be analyzed with these SAS procedures.

SAS System for Regression, 1986 Edition represents an introduction to regression analysis as performed by the SAS System. In addition to the most current programming conventions for both Version 5 and Version 6 of the SAS System, this volume contains information about new features and capabilities of several SAS procedures. Here is a summary of the information contained in each chapter:

Chapter 1, "Regression Concepts," covers three topics: the terminology and notation used in regression analysis; an overview of matrix notation and PROC IML; and regression procedures available in the SAS System.

In Chapter 2, "Using PROC REG," regression analysis is introduced by using PROC REG with a single independent variable. The available statistics that appear on the printed output are discussed in detail. Next, a regression is performed using the same data but with several independent variables. The MODEL statement options P, CLM, and CLI are used and explained. The NOINT option, which is very controversial, is used and discussed. This option is also misused in an example to point out the dangers of forcing the regression response to pass through the origin when this situation is unreasonable or unlikely. Output data sets are created using the statistics generated with PROC REG.

Chapter 3, "Observations," discusses ways to identify the influence (or leverage) of specific observations. Outliers, or those observations that do not appear to fit the model, can bias parameter estimates and make your regression analysis less useful. Studentized residuals are used to identify large residuals. Examples are shown using residuals that are plotted using PROC PLOT and PROC GPLOT.

Chapter 4, "Multicollinearity: Detection and Remedial Measures," discusses the existence of multicollinearity, the high degree of correlation among several independent variables, and measures you can use to detect multicollinearity and then eliminate it. Variance inflation factors are used to determine the variables involved. Multivariate techniques are used to study the structure of multicollinearity.

Chapter 5, "Polynomial Models," gives examples of linear regression methods to estimate parameters of models that cannot be described by straight lines. The discussion of polynomial models, a model in which the dependent variable is related to functions of the powers of one or more independent variables, begins by using one independent variable. Then examples are given using several variables. Response surface plots are used to illustrate the nature of the estimated response curve.

Chapter 6, "Special Models," treats relationships that are not linear, such as log-linear models, spline functions, and restricted linear models. Plots are used in conjunction with the regression analysis in order to determine the pattern of the residuals. Cautionary measures and the results of the analysis are discussed along with PROC NLIN.

The appendix provides a brief discussion of statistical terms used in regression analysis.

The discussions and examples in this book should facilitate your understanding of regression analysis as performed by the SAS System and help you in building more powerful regression models.

If you have any comments or suggestions about SAS software or the *SAS System for Regression*, please write your ideas on a photocopy of the "Your Turn" page at the back of this book.

1. Regression Concepts

1.1 STATISTICAL BACKGROUND

Multiple linear regression analysis is an inference procedure using the model

$$y = \beta_0 + \beta_1 x_1 + \beta_2 x_2 + \ldots + \beta_m x_m + \varepsilon$$

relating the behavior of a dependent variable y to a linear function of the set of independent variables $x_1, x_2, \ldots, x_m$. The β_js are the parameters that specify the nature of the relationship, and ε is the random error, a term that takes into account the fact that the model does not exactly describe the behavior of the data. Although you should have a basic understanding of regression analysis, this chapter begins with a review of regression principles, terminology, notation, and procedures.

1.1.1 Terminology and Notation

The principle of *least squares* is applied to a set of n observed values of y and the associated x_j to obtain estimates $\hat{\beta}_0, \hat{\beta}_1, \ldots, \hat{\beta}_m$ of the respective parameters $\beta_0, \beta_1, \ldots, \beta_m$. These estimates are then used to construct the fitted model, or estimating equation

$$\hat{y} = \hat{\beta}_0 + \hat{\beta}_1 x_1 + \ldots + \hat{\beta}_m x_m \quad .$$

Many regression computations are illustrated conveniently in matrix notation. Let y_i, x_{ij}, and ε_i denote the values of y, x_j, and ε, respectively, in the ith observation. The **Y** vector, the **X** matrix, and the ε vector can be defined as follows:

$$\mathbf{Y} = \begin{bmatrix} y_1 \\ \cdot \\ \cdot \\ \cdot \\ y_n \end{bmatrix}, \mathbf{X} = \begin{bmatrix} 1 & x_{11} & \ldots & x_{1m} \\ \cdot & \cdot & & \cdot \\ \cdot & \cdot & & \cdot \\ \cdot & \cdot & & \cdot \\ 1 & x_{n1} & \ldots & x_{nm} \end{bmatrix}, \varepsilon = \begin{bmatrix} \varepsilon_1 \\ \cdot \\ \cdot \\ \cdot \\ \varepsilon_n \end{bmatrix} .$$

Then the model in matrix notation is

$$\mathbf{Y} = \mathbf{X}\boldsymbol{\beta} + \boldsymbol{\varepsilon}$$

where $\boldsymbol{\beta}' = (\beta_0, \beta_1, \ldots, \beta_m)$ is the parameter vector.
The vector of least-squares estimates

$$\hat{\boldsymbol{\beta}}' = (\hat{\beta}_0, \hat{\beta}_1, \ldots, \hat{\beta}_m)$$

is obtained by solving the set of normal equations (NE)

$$\mathbf{X}'\mathbf{X}\boldsymbol{\beta} = \mathbf{X}'\mathbf{Y} \quad .$$

Assuming that $\mathbf{X}'\mathbf{X}$ is of full rank or nonsingular, there is a unique solution to the NEs given by

$$\hat{\boldsymbol{\beta}} = (\mathbf{X}'\mathbf{X})^{-1}\mathbf{X}'\mathbf{Y} \quad .$$

The matrix $(\mathbf{X}'\mathbf{X})^{-1}$ is very useful in regression analysis and is often denoted as follows:

$$(\mathbf{X}'\mathbf{X})^{-1} = \mathbf{C} = \begin{bmatrix} c_{00} & c_{01} & \cdots & c_{0m} \\ c_{10} & c_{11} & \cdots & c_{1m} \\ \cdot & \cdot & & \cdot \\ \cdot & \cdot & & \cdot \\ \cdot & \cdot & & \cdot \\ c_{m0} & c_{m1} & \cdots & c_{mm} \end{bmatrix} \quad .$$

1.1.2 Partitioning the Sums of Squares

A basic identity results from least squares, specifically,

$$\Sigma(y - \bar{y})^2 = \Sigma(\bar{y} - \hat{y})^2 + \Sigma(y - \hat{y})^2 \quad .$$

This identity shows that the total sum of squared deviations from the mean, $\Sigma(y-\bar{y})^2$, is equal to the sum of squared differences between the mean and the predicted values, $\Sigma(\bar{y}-\hat{y})^2$, plus the sum of squared deviations from the observed ys to the regression line, $\Sigma(y-\hat{y})^2$. These two parts are called the sum of squares due to regression (or model) and the residual (or error) sum of squares. Thus,

$$\text{Total SS} = \text{Model SS} + \text{Residual SS} \quad .$$

Total SS always has the same value for a given set of data, regardless of the model that is fitted; however, partitioning into Model SS and Residual SS depends on the model. Generally, the addition of a new x variable to a model increases the Model SS and, correspondingly, reduces the Residual SS. The residual, or error, sum of squares is computed as follows:

$$\begin{aligned} \text{Error SS} &= \mathbf{Y}'(\mathbf{I} - \mathbf{X}(\mathbf{X}'\mathbf{X})^{-1}\mathbf{X}')\mathbf{Y} \\ &= \mathbf{Y}'\mathbf{Y} - \mathbf{Y}'\mathbf{X}(\mathbf{X}'\mathbf{X})^{-1}\mathbf{X}'\mathbf{Y} \\ &= \mathbf{Y}'\mathbf{Y} - \hat{\boldsymbol{\beta}}'\mathbf{X}'\mathbf{Y} \quad . \end{aligned}$$

The error mean square

$$s^2 = \text{MSE} = (\text{Error SS}) / (n - m - 1)$$

is an unbiased estimate of σ^2, the variance of the ε_is.

Sums of squares, including the different sums of squares computed by any regression procedure such as PROC REG and PROC GLM, can be expressed conceptually as the difference between the regression sums of squares for two models, called complete and reduced models, respectively. This approach relates a given SS to the comparison of two regression models.

For example, denote as SS_1 the regression sum of squares for a complete model with m=5 variables:

$$y = \beta_0 + \beta_1 x_1 + \beta_2 x_2 + \beta_3 x_3 + \beta_4 x_4 + \beta_5 x_5 + \varepsilon \quad .$$

Denote the regression sum of squares for a reduced model not containing x_4 and x_5 as SS_2:

$$y = \beta_0 + \beta_1 x_1 + \beta_2 x_2 + \beta_3 x_3 + \varepsilon \quad .$$

Reduction notation can be used to represent the difference between regression sums of squares for the two models:

$$R(\beta_4, \beta_5 \mid \beta_0, \beta_1, \beta_2, \beta_3) = \text{Model SS}_1 - \text{Model SS}_2 \quad .$$

The difference or reduction in error $R(\beta_4, \beta_5 \mid \beta_0, \beta_1, \beta_2, \beta_3)$ indicates the increase in regression sums of squares due to the addition of β_4 and β_5 to the reduced model. It follows that

$$R(\beta_4, \beta_5 \mid \beta_0, \beta_1, \beta_2, \beta_3) = \text{Residual SS}_2 - \text{Residual SS}_1 \quad .$$

The expression

$$R(\beta_4, \beta_5 \mid \beta_0, \beta_1, \beta_2, \beta_3)$$

is also commonly referred to in the following ways:

1. the sums of squares due to β_4 and β_5 (or x_4 and x_5) adjusted for β_0, β_1, β_2, β_3 (or the intercept and x_1, x_2, x_3)
2. the sums of squares due to fitting x_4 and x_5 after fitting the intercept and x_1, x_2, x_3
3. the effects of x_4 and x_5 above and beyond or partialing the effects of the intercept and x_1, x_2, x_3.

1.1.3 Hypothesis Testing

Inferences about model parameters are highly dependent on the other parameters in the model under consideration; thus, in hypothesis testing, it is important to emphasize the parameters for which inferences have been adjusted. For example, $R(\beta_3 \mid \beta_0, \beta_1, \beta_2)$ and $R(\beta_3 \mid \beta_0, \beta_1)$ may measure entirely different concepts. Similarly, a test of H_0: β_3=0 versus H_1: $\beta_3 \neq 0$ may have one result for the model

$$y = \beta_0 + \beta_1 x_1 + \beta_3 x_3 + \varepsilon$$

and another for the model

$$y = \beta_0 + \beta_1 x_1 + \beta_2 x_2 + \beta_3 x_3 + \varepsilon \quad .$$

Differences reflect actual physical dependencies among parameters in the model rather than inconsistencies in statistical methodology.

Statistical inferences can also be made in terms of linear functions of the parameters of the form

$$H_0\text{: } \mathbf{L}\beta\text{: } \ell_0\beta_0 + \ell_1\beta_1 + \ldots + \ell_m\beta_m = 0$$

where the ℓ_is are arbitrary constants chosen to correspond to a specified hypothesis. Such functions are estimated by the corresponding linear function

$$\mathbf{L}\hat{\beta} = \ell_0\hat{\beta}_0 + \ell_1\hat{\beta}_1 + \ldots + \ell_m\hat{\beta}_m$$

of the least-squares estimates $\hat{\beta}$. The variance of $\mathbf{L}\hat{\beta}$ is

$$V(\mathbf{L}\hat{\beta}) = (\mathbf{L}(\mathbf{X}'\mathbf{X})^{-1}\mathbf{L}')\sigma^2 \quad .$$

Then a t test or F test is used to test H_0: $(\mathbf{L}\beta)=0$. The denominator is usually the residual mean square (MSE). Because the variance of the estimated function is based on statistics computed for the entire model, the test of the hypothesis is made in the presence of all model parameters. Confidence intervals can be constructed to correspond to these tests, which can be generalized to simultaneous tests of several linear functions.

Simultaneous inference about a set of linear functions $\mathbf{L}_1\beta, \ldots, \mathbf{L}_k\beta$ is performed in a related manner. For notational convenience, let $\mathbf{L}$ denote the matrix whose rows are $\mathbf{L}_1, \ldots, \mathbf{L}_k$:

$$\mathbf{L} = \begin{bmatrix} \mathbf{L}_1 \\ \cdot \\ \cdot \\ \cdot \\ \mathbf{L}_k \end{bmatrix} \quad .$$

Then the sum of squares

$$SS(\mathbf{L}\beta = 0) = (\mathbf{L}\hat{\beta})'(\mathbf{L}(\mathbf{X}'\mathbf{X})^{-1}\mathbf{L}')^{-1}(\mathbf{L}\hat{\beta})$$

is associated with the null hypothesis

$$H_0\text{: } \mathbf{L}_1\beta = \ldots = \mathbf{L}_k\beta = 0 \quad .$$

A test of H_0 is provided by the F statistic

$$F = (SS(\mathbf{L}\beta = 0) / k) / MSE \quad .$$

Three common types of statistical inferences are as follows:

1. a test that all parameters ($\beta_1, \beta_2, \ldots, \beta_m$) are zero. The test compares the fit of the complete model to that using only the mean:

$$F = (\text{Model SS} / m) / \text{MSE}$$

where

$$\text{Model SS} = R(\beta_1, \beta_2, \ldots, \beta_m \mid \beta_0) \quad .^*$$

The F statistic has (m, $n-m-1$) degrees of freedom.

2. a test that the parameters in a subset are zero. The problem is to compare the fit of the complete model

$$y = \beta_0 + \beta_1 x_1 + \ldots + \beta_g x_g + \beta_{g+1} x_{g+1} + \ldots + \beta_m x_m + \varepsilon$$

to the fit of the reduced model

$$y = \beta_0 + \beta_1 x_1 + \ldots + \beta_g x_g + \varepsilon \quad .$$

An F statistic is used to perform the test

$$F = (R(\beta_{g+1}, \ldots, \beta_m \mid \beta_0, \beta_1, \ldots, \beta_g) / (m - g)) / \text{MSE} \quad .$$

Note that an arbitrary reordering of variables produces a test for any desired subset of parameters. If the subset contains only one parameter, β_m, the test is

$$\begin{aligned} F &= (R(\beta_m \mid \beta_0, \beta_1, \ldots, \beta_{m-1}) / 1) / \text{MSE} \\ &= (\text{partial SS due to } \beta_m) / \text{MSE} \end{aligned}$$

which is equivalent to the t test

$$t = \hat{\beta}_m / s_{\hat{\beta}m} = \hat{\beta}_m / \sqrt{c_{mm}\text{MSE}} \quad .$$

The corresponding $(1-\alpha)$ confidence interval about β_m is

$$\hat{\beta}_m \pm t_{\alpha/2} \sqrt{c_{mm}\text{MSE}} \quad .$$

3. estimation of a subpopulation mean corresponding to a specific x. For a given set of x values described by a vector $\mathbf{x}$, the estimated population mean is

$$E(y_{\mathbf{x}}) = \hat{\beta}_0 + \hat{\beta}_1 \mathbf{x}_1 + \ldots + \hat{\beta}_m \mathbf{x}_m = \mathbf{x}'\hat{\beta} \quad .$$

The vector $\mathbf{x}$ is constant; hence, the variance of $E(y_{\mathbf{x}})$ is

$$V(E(y_{\mathbf{x}})) = \mathbf{x}'(\mathbf{X}'\mathbf{X})^{-1}\mathbf{x}(\text{MSE}) \quad .$$

This equation is useful for computing confidence intervals.

---

* $R(\beta_0, \beta_1, \ldots, \beta_m)$ is rarely used. For more information, see the NOINT option in section **2.4.5**.

A related inference concerns a future single value of y corresponding to a specified x. The relevant variance estimate is

$$V(y_{\mathbf{x}}) = (1 + \mathbf{x}'(\mathbf{X}'\mathbf{X})^{-1}\mathbf{x})\text{MSE} \quad .$$

1.1.4 Using the Generalized Inverse

Many applications of regression procedures involve an **X′X** matrix that is not of full rank and has no unique inverse. PROC GLM, as well as PROC REG, computes a generalized inverse $(\mathbf{X}'\mathbf{X})^{-}$ and uses it to compute a regression estimate

$$\mathbf{b} = (\mathbf{X}'\mathbf{X})^{-}\mathbf{X}'\mathbf{Y} \quad .$$

A generalized inverse of a matrix **A** is any matrix **G** such that **AGA**=**A**. Note that this also identifies the inverse of a full-rank matrix.

If **X′X** is not of full rank, then an infinite number of generalized inverses exist. Different generalized inverses lead to different solutions to the normal equations that will have different expected values; that is, $E(\mathbf{b})=(\mathbf{X}'\mathbf{X})^{-}\mathbf{X}'\mathbf{X}\boldsymbol{\beta}$ depends on the particular generalized inverse used to obtain **b**. Therefore, it is important to understand what is being estimated by the solution.

Fortunately, not all computations in regression analysis depend on the particular solution obtained. For example, the error sum of squares is invariant with respect to $(\mathbf{X}'\mathbf{X})^{-}$ and is given by

$$\text{SSE} = \mathbf{Y}'(\mathbf{I} - \mathbf{X}(\mathbf{X}'\mathbf{X})^{-}\mathbf{X}')\mathbf{Y} \quad .$$

Hence, the model sum of squares also does not depend on the particular generalized inverse obtained.

The generalized inverse has played a major role in the presentation of the theory of linear statistical models, notably in the work of Graybill (1976) and Searle (1971). In a theoretical setting it is often possible, and even desirable, to avoid specifying a particular generalized inverse. To apply the generalized inverse to statistical data using computer programs, a generalized inverse must actually be calculated. Therefore, it is necessary to declare the specific generalized inverse being computed. For example, consider an **X′X** matrix of rank k that can be partitioned as

$$\mathbf{X}'\mathbf{X} = \begin{bmatrix} \mathbf{A}_{11} & \mathbf{A}_{12} \\ \mathbf{A}_{21} & \mathbf{A}_{22} \end{bmatrix}$$

where $\mathbf{A}_{11}$ is k x k and of rank k. Then $\mathbf{A}_{11}^{-1}$ exists, and a generalized inverse of **X′X** is

$$(\mathbf{X}'\mathbf{X})^{-} = \begin{bmatrix} \mathbf{A}_{11}^{-1} & \boldsymbol{\varphi}_{12} \\ \boldsymbol{\varphi}_{21} & \boldsymbol{\varphi}_{22} \end{bmatrix}$$

where each $\boldsymbol{\varphi}_{ij}$ is a matrix of zeros of the same dimension as $\mathbf{A}_{ij}$.

This approach to obtaining a generalized inverse, the method used by PROC GLM and PROC REG, can be extended indefinitely by partitioning a singular matrix into several sets of matrices as illustrated above. Note that the resulting solution to the normal equations, $\mathbf{b}=(\mathbf{X}'\mathbf{X})^{-}\mathbf{X}'\mathbf{Y}$, has zeros in the positions corresponding to the rows filled with zeros in $(\mathbf{X'X})^-$. This is the solution printed by these procedures, and it is regarded as providing a biased estimate of β.

However, because **b** is not unique, a linear function, **Lb**, and its variance are generally not unique either. However, a class of linear functions called *estimable functions* exists, and they have the following properties:

1. The vector **L** is a linear combination of rows of **X**.
2. **Lb** and its variance are invariant through all possible generalized inverses. In other words, **Lb** is unique and is an unbiased estimate of $\mathbf{L}\beta$.

Analogous to the full-rank case, the variance of an estimable function **Lb** is given by

$$V(\mathbf{Lb}) = (\mathbf{L}(\mathbf{X'X})^- \mathbf{L'})\sigma^2 \quad .$$

This expression is used for statistical inference. For example, a test of H_0: $\mathbf{L}\beta=0$ is given by the *t* test

$$t = \mathbf{Lb} \; / \; \sqrt{(\mathbf{L}(\mathbf{X'X})^- \mathbf{L'})\mathrm{MSE}} \quad .$$

Simultaneous inferences on a set of estimable functions are performed in an analogous manner.

1.2 PERFORMING A REGRESSION WITH PROC IML

As you will see in section **1.3** and in greater detail in subsequent chapters, the SAS System provides a flexible array of procedures for performing regression analyses. You can also perform these analyses by direct application of the matrix formulas presented in the previous section using PROC IML. PROC IML is most frequently used for the custom programming of methods too specialized or too new to be packaged into the standard regression procedures. It is also useful as an instructional tool for illustrating linear model and other methodologies.

The following example illustrates a regression analysis performed by PROC IML. This example is not intended to serve as a tutorial in the use of PROC IML. If you need more information on PROC IML, refer either to the *SAS/IML User's Guide, Version 5 Edition* or the *SAS/IML Guide for Personal Computers, Version 6 Edition.*

The example for this section is also used in Chapter 2 to illustrate PROC REG. The data set is described, and the data are presented in section **2.1**. For this presentation, the variable CPM is the dependent variable *y*, and UTL, SPA, and ALF are the independent variables x_1, x_2, and x_3, respectively. Comment statements are used in the SAS program to explain the individual steps in the analysis.

```
*INVOKE PROC IML. CREATE THE X AND Y MATRICES FROM THE
 DATA SET.;

PROC IML;
   USE AIR;
   READ ALL VAR {"UTL" "SPA" "ALF"} INTO X;
   READ ALL VAR {"CPM"} INTO Y;

*DEFINE THE NUMBER OF OBSERVATIONS (N) AND THE NUMBER OF
 VARIABLES (M) AS THE NUMBER OF ROWS AND COLUMNS OF X. ADD A
 COLUMN OF ONES, FOR THE INTERCEPT VARIABLE, TO THE X
 MATRIX.;
```

```
   N=NROW(X);
   M=NCOL(X);
   X=J(N,1,1)||X;

*COMPUTE C, THE INVERSE OF X'X AND THE VECTOR OF COEFFICIENT
 ESTIMATES BHAT.;

   C=INV(X'*X);
   BHAT=C*X'*Y;

*COMPUTE SSE, THE RESIDUAL SUM OF SQUARES, AND MSE, THE
 RESIDUAL MEAN SQUARE (VARIANCE ESTIMATE).;

   SSE=Y'*Y-BHAT'*X'*Y;
   DFE=N-M-1;
   MSE=SSE / DFE;

*THE TEST FOR THE MODEL CAN BE RESTATED AS A TEST FOR THE
 LINEAR FUNCTION LB=0, WHERE L IS THE MATRIX.;

   L={0 1 0 0,
      0 0 1 0,
      0 0 0 1};

*COMPUTE SSMODEL FOR THIS HYPOTHESIS AND THE CORRESPONDING
 F RATIO.;

   SSMODEL=(L*BHAT)'*INV(L*C*L')*(L*BHAT);
   F=(SSMODEL / M) / MSE;

*COMPUTE SEB, THE VECTOR OF STANDARD ERRORS OF THE ESTIMATED
 COEFFICIENTS, AND THE MATRIX T CONTAINING THE T STATISTIC
 FOR TESTING THAT EACH COEFFICIENT IS ZERO. FINALLY, CREATE
 THE MATRIX STATS, WHICH CONTAINS AS ITS COLUMNS THE
 COEFFICIENT ESTIMATES, THEIR STANDARD ERRORS, AND THE
 T STATISTICS, RESPECTIVELY.;

  SEB=SQRT(VECDIAG(C)#MSE);
  T=BHAT / SEB;
  PROBT=1-PROBF(T#T,1,DFE);
  STATS=BHAT||SEB||T||PROBT;

*COMPUTE THE PREDICTED VALUES (YHAT) AND THE RESIDUAL VALUES
 (RESID).  CONSTRUCT A MATRIX (OBS) CONTAINING AS ITS
 COLUMNS THE ACTUAL, PREDICTED, AND RESIDUAL VALUES,
 RESPECTIVELY.;

   YHAT=X*BHAT;
   RESID=Y-YHAT;
   OBS=Y||YHAT||RESID;
```

```
*PRINT THE MATRICES CONTAINING THE DESIRED RESULTS.;

   PRINT SSMODEL,,,M,,,MSE,,,DFE,,,STATS
   (|COLNAME={"BHAT","SEB","T","PROBT"}|),,,OBS
   (|FORMAT=8.3 COLNAME={"Y""YHAT""RESID"}|);
```

The results of this sample program are shown in **Output 1.1**.

When you use PROC IML, all results are in the form of matrices. Each matrix is identified by its name, and its elements are identified by row and column indices. You may find it necessary to refer to the program to identify specific elements.

Output 1.1 Output Produced by PROC IML

```
              SSMODEL      COL1

              ROW1       6.0471

               M          COL1

               ROW1   3.0000

               MSE        COL1

               ROW1   0.1683

               F          COL1

               ROW1  11.9745

               DFE        COL1

               ROW1  29.0000


  STATS     BHAT      SEB        T         PROBT

  ROW1    7.7215   0.7976   9.6807   1.4E-10
  ROW2   -0.1385   0.0530  -2.6135    0.0141
  ROW3   -3.5036   0.9642  -3.6337  .0010706
  ROW4   -6.2031   1.2489  -4.9669   2.8E-05


      OBS        Y         YHAT      RESID

      ROW1     2.258     2.484    -0.226
      ROW2     2.275     2.136     0.139
      ROW3     2.341     3.398    -1.057
      ROW4     2.357     2.229     0.128
      ROW5     2.363     2.453    -0.090
      ROW6     2.404     3.013    -0.609
      ROW7     2.425     2.430    -0.005
      ROW8     2.711     2.769    -0.058
      ROW9     2.743     3.268    -0.525
      ROW10    2.780     2.760     0.020
      ROW11    2.833     2.814     0.019
      ROW12    2.846     3.359    -0.513
      ROW13    2.906     3.148    -0.242
      ROW14    2.954     3.066    -0.112
      ROW15    2.962     3.102    -0.140
      ROW16    2.971     3.028    -0.057
      ROW17    3.044     3.451    -0.407
      ROW18    3.096     2.892     0.204
      ROW19    3.140     3.187    -0.047
      ROW20    3.306     3.587    -0.281
      ROW21    3.306     2.812     0.494
      ROW22    3.311     3.405    -0.094
      ROW23    3.313     3.397    -0.084
```

(continued on next page)

(continued from previous page)

```
ROW24    3.392    3.469   -0.077
ROW25    3.437    3.361    0.076
ROW26    3.462    3.328    0.134
ROW27    3.527    3.084    0.443
ROW28    3.689    3.656    0.033
ROW29    3.760    3.427    0.333
ROW30    3.856    3.168    0.688
ROW31    3.959    3.505    0.454
ROW32    4.024    3.161    0.863
ROW33    4.737    4.142    0.595
```

The results of this analysis are discussed thoroughly in Chapter 2; therefore, in this section only the results that can be compared with those from PROC REG (shown in **Output 2.3**) are identified.

The statistics in the output are not in the order that they were computed; instead, they are presented in the order found in the PROC REG output. The first five entries are scalars representing the statistics for the entire model. The first two entries are the sum of squares and degrees of freedom for the model; the next two are the residual mean square and the corresponding degrees of freedom; the last entry is the F ratio to test for the existence of the model.

The matrix STATS contains the information on the parameter estimates. Rows correspond to parameters (intercept and coefficients for x_1, x_2, and x_3, respectively), and columns correspond to the different statistics. The first column contains the coefficient estimates (from matrix BHAT), the second contains the standard errors of the estimates (from matrix SEB), and the third contains the t statistics (from matrix T).

The matrix OBS contains the information on observations. The rows correspond to the observations. Column one contains the original y values (matrix Y), column two contains the predicted values (from matrix YHAT), and column three contains the residuals (from matrix RESID).

The results achieved by using PROC IML agree with those from PROC REG, as shown in **Output 2.3**. Because PROC IML is most frequently used for the custom programming of new or specialized methods, the standard regression procedures are more efficient with respect to both programming time and computing time. For this reason, you should try to use these procedures whenever possible. In addition, the output produced with the standard regression procedures is designed to present analysis results more clearly than do the printed matrices produced with PROC IML. See section **1.3** for an overview of standard regression procedures.

1.3 REGRESSION WITH THE SAS SYSTEM

This section presents a brief overview of the various procedures the SAS System makes available for performing regression analyses.

PROC AUTOREG

implements the appropriate regression analysis when autocorrelation exists among the errors.

Software product: SAS/ETS® software

Documentation: *SAS/ETS® User's Guide, Version 5 Edition*

PROC GLM
can also be used for any multiple regression analysis; however, it is primarily designed for estimating parameters of analysis-of-variance and covariance models (see Freund and Littell 1986). Because of its greater flexibility, PROC GLM uses more computer resources than PROC REG and is usually not recommended for regression analyses. It also does not have a number of the diagnostic abilities of PROC REG.

Software product: Base SAS® software

Documentation: *SAS® User's Guide: Statistics, Version 5 Edition*

Software product: SAS/STAT™ software

Documentation: *SAS/STAT™ Guide for Personal Computers, Version 6 Edition*

PROC NLIN
implements several estimation procedures for nonlinear models, which may include restrictions on parameter values.

Software product: Base SAS® software

Documentation: *SAS® User's Guide: Statistics, Version 5 Edition*

Software product: SAS/STAT™ software

Documentation: *SAS/STAT™ Guide for Personal Computers, Version 6 Edition, Volume 2*

PROC ORTHOREG
performs regression using the Gentleman-Givens method, which is numerically stable while still allowing you to avoid storing the entire data matrix in memory. For ill-conditioned data, ORTHOREG uses very specialized calculations that can be more accurate.

Software product: SAS/STAT™ software

Documentation: *SAS/STAT™ Guide for Personal Computers, Version 6 Edition, Volume 2**

PROC REG
is the principal procedure of the SAS System for performing regression analyses. In addition to providing the usual estimates and inference statistics associated with regression, PROC REG provides many options for special estimates, outlier and specification error detection (row diagnostics), collinearity statistics (column diagnostics), and tests on linear functions of parameter estimates. It can perform restricted least-squares estimation and multivariate tests,

* PROC ORTHOREG was previously documented in SAS Technical Report P-161.

and it can detect and define exact collinearities. In addition PROC REG can produce SAS data sets containing most of the estimates and test statistics produced by the procedure.

Software product: Base SAS® software

Documentation: *SAS® User's Guide: Statistics, Version 5 Edition*

Software product: SAS/STAT™ software

Documentation: *SAS/STAT™ Guide for Personal Computers, Version 6 Edition*

PROC RSREG
implements a quadratic response surface regression, including a search for optimum response. It can also produce a SAS data set that can be used to construct response surface plots.

Software product: Base SAS® software

Documentation: *SAS® User's Guide: Statistics, Version 5 Edition*

Software product: SAS/STAT™ software

Documentation: *SAS/STAT™ Guide for Personal Computers, Version 6 Edition, Volume 2*

PROC STEPWISE and PROC RSQUARE
are used for variable selection (model building). PROC STEPWISE implements several one-at-a-time deletion, addition, or swapping strategies for selecting subsets of independent variables. PROC RSQUARE computes the summary statistics for all possible subsets of variables.

Software product: Base SAS® software

Documentation: *SAS® User's Guide: Statistics, Version 5 Edition*

Software product: SAS/STAT™ software

Documentation: *SAS/STAT™ Guide for Personal Computers, Version 6 Edition, Volume 2*

Other specialized procedures for example, PROC LAV and PROC RIDGEREG, are documented in the *SUGI Supplemental Library User's Guide*, and other procedures, which are primarily useful for econometric analyses, are documented in the *SAS/ETS User's Guide.*

1.4 SAS SOFTWARE PROCEDURES DISCUSSED IN THIS BOOK

Examples showing how to use the procedures discussed in section **1.3** are presented throughout this book. Below is a brief description of the chapters in this book. Each description mentions the particular SAS software procedures examined in each chapter.

Chapter 2, "Using PROC REG," provides the information for using PROC REG to perform a regression analysis, including parameter estimates, hypothesis tests, inferences for prediction, and instructions for creating data sets containing computed statistics.

Chapter 3, "Observations," shows how PROC REG can be used to detect failures in assumptions, including outliers, specification errors, heterogeneous variances, and autocorrelated errors. Remedial methods include weighted regression (with PROC REG) and the use of PROC LAV and PROC AUTOREG.

Chapter 4, "Multicollinearity: Detection and Remedial Measures," presents multicollinearity detection statistics provided by PROC REG. Remedial measures include variable selection with PROC STEPWISE and PROC RSQUARE, and principal component regression (using PROC FACTOR and PROC REG).

Chapter 5, "Polynomial Models," presents polynomial regression using PROC REG and PROC RSREG and also shows how to construct plots that may be useful for the search for an adequate model.

Chapter 6, "Special Models," presents the implementation of some special models, including log-linear models, strictly nonlinear models using PROC NLIN, spline models using PROC REG and PROC NLIN, and the use of indicator variables.

2. Using PROC REG

2.1 INTRODUCTION

As indicated in section **1.3**, PROC REG is the primary SAS software procedure for performing the computations for a statistical analysis of data based on a linear regression model. The basic set of statements for implementing such an analysis is as follows:

```
PROC REG;
   MODEL list of dependent variables = list of independent
         variables / model options;
```

This chapter provides instructions in the use of PROC REG for performing a regression analysis. Subsequent chapters deal with special options and other procedures for more specialized analyses and models. The data for the example used in this chapter concern factors considered to be influential in determining the cost of providing air service. This example develops a model for estimating the cost-per-passenger mile in order to isolate the major factors in determining that cost. The data are taken from a Civil Aeronautics Board report "Aircraft Operation Costs and Performance Report," August, 1972. These data are used to create a SAS data set, AIR. The following variables are considered:

CPM is cost-per-passenger mile (cents).

UTL is average hours per day use of aircraft.

ASL is average length of nonstop legs of flights (1000 miles).

SPA is average number of seats per aircraft (100 seats).

ALF is average load factor (percentage of seats occupied by passengers).

The data represent thirty-three US airlines with average nonstop lengths of flights greater than 800 miles. An additional indicator variable, TYPE, has been constructed. This variable has value zero for airlines with ASL$<$1200 miles and unity for airlines with ASL$\geq$1200 miles. This variable will be used in section **2.7.3**. The data are given in **Output 2.1**.

```
DATA AIR;
   INPUT UTL 1-4 2 ASL 5-8 3 SPA 9-12 4 AVS 13-15 2
         ALF 16-18 3 CPM 19-22 3 CSM 23-26 3;
   TYPE=0;
   IF ASL>1.20 THEN TYPE=1;
   CARDS;
78717901375468591225813333
95025153546483488227511111
79113501920442412234109655
13303607339050139723570935
more data lines
;
PROC SORT;
   BY ALF;
PROC PRINT;
   VAR ALF UTL ASL SPA TYPE CPM;
```

Output 2.1 Airline Data Using Indicator Variable TYPE=

OBS	ALF	UTL	ASL	SPA	TYPE	CPM
1	0.287	8.09	1.528	0.3522	1	3.306
2	0.349	9.56	2.189	0.3279	1	3.527
3	0.362	10.80	1.518	0.1356	1	3.959
4	0.378	5.65	0.821	0.1290	0	4.737
5	0.381	10.20	1.692	0.3007	1	3.096
6	0.394	7.94	0.949	0.1488	0	3.689
7	0.397	13.30	3.607	0.3390	1	2.357
8	0.400	8.42	1.495	0.3597	1	2.833
9	0.405	9.57	0.863	0.1390	0	3.313
10	0.409	9.00	0.845	0.1390	0	3.044
11	0.410	9.62	0.840	0.1390	0	2.846
12	0.412	7.91	1.350	0.1920	1	2.341
13	0.417	8.83	2.377	0.3287	1	2.780
14	0.422	8.35	1.031	0.1365	0	3.392
15	0.425	10.60	2.780	0.1282	1	3.856
16	0.426	7.52	0.975	0.2025	0	3.462
17	0.434	8.36	1.912	0.3148	1	2.711
18	0.439	8.43	1.584	0.1607	1	2.743
19	0.452	7.55	1.164	0.1270	0	3.760
20	0.455	7.70	1.236	0.1221	1	3.311
21	0.466	9.38	1.123	0.1481	0	2.404
22	0.476	8.91	0.961	0.1236	0	2.962
23	0.476	7.27	1.416	0.1145	1	3.437
24	0.478	8.71	1.392	0.1148	1	2.906
25	0.486	8.29	0.877	0.1060	0	3.140
26	0.488	9.50	2.515	0.3546	1	2.275
27	0.495	8.44	0.871	0.1186	0	2.954
28	0.504	9.47	1.408	0.1345	1	3.306
29	0.535	10.80	1.576	0.1361	1	2.425
30	0.539	6.84	1.008	0.1150	0	2.971

(continued on next page)

(continued from previous page)

31	0.541	6.31	0.823	0.0943	0	4.024
32	0.582	8.48	1.963	0.1381	1	2.363
33	0.591	7.87	1.790	0.1375	1	2.258

2.2 A MODEL WITH ONE INDEPENDENT VARIABLE

In the SAS program below, PROC REG is used to perform a regression analysis for a model with a single independent, or regressor, variable. For example, for a regression of a dependent variable Y on a single independent variable X, the required statements are as follows:

```
PROC REG;
   MODEL Y=X;
```

This model is illustrated by using only the variable ALF, the average load factor, to estimate the cost-per-passenger mile, CPM. In terms of the example, the model is as follows:

$$CPM = \beta_0 + \beta_1(ALF) + \varepsilon \quad .$$

In this model, β_1 is the effect on cost of a one percent increase in the load factor. Normally, you would expect this coefficient to be negative. The coefficient β_0, the intercept, is the cost-per-passenger mile if the load factor is zero. Obviously, this is not a useful value, but the term is needed to specify fully the regression line. The term ε, representing the random errors, accounts for variation in costs due to other factors. In the case of a regression with one independent variable, plotting the observed variables is sometimes instructive.

The plot shows the expected tendency for lower costs (CPM) with higher load factors (ALF). However, the relationship is not very strong because of the other obvious cost factors that you will see in section **2.3**.

The following SAS statements are used to plot CPM against ALF:

```
PROC PLOT DATA=AIR;
   PLOT CPM*ALF;
```

A plot of the values of CPM and ALF is presented in **Output 2.2**. The SAS statements required for this model are as follows:

```
PROC REG DATA=AIR;
   MODEL CPM=ALF;
```

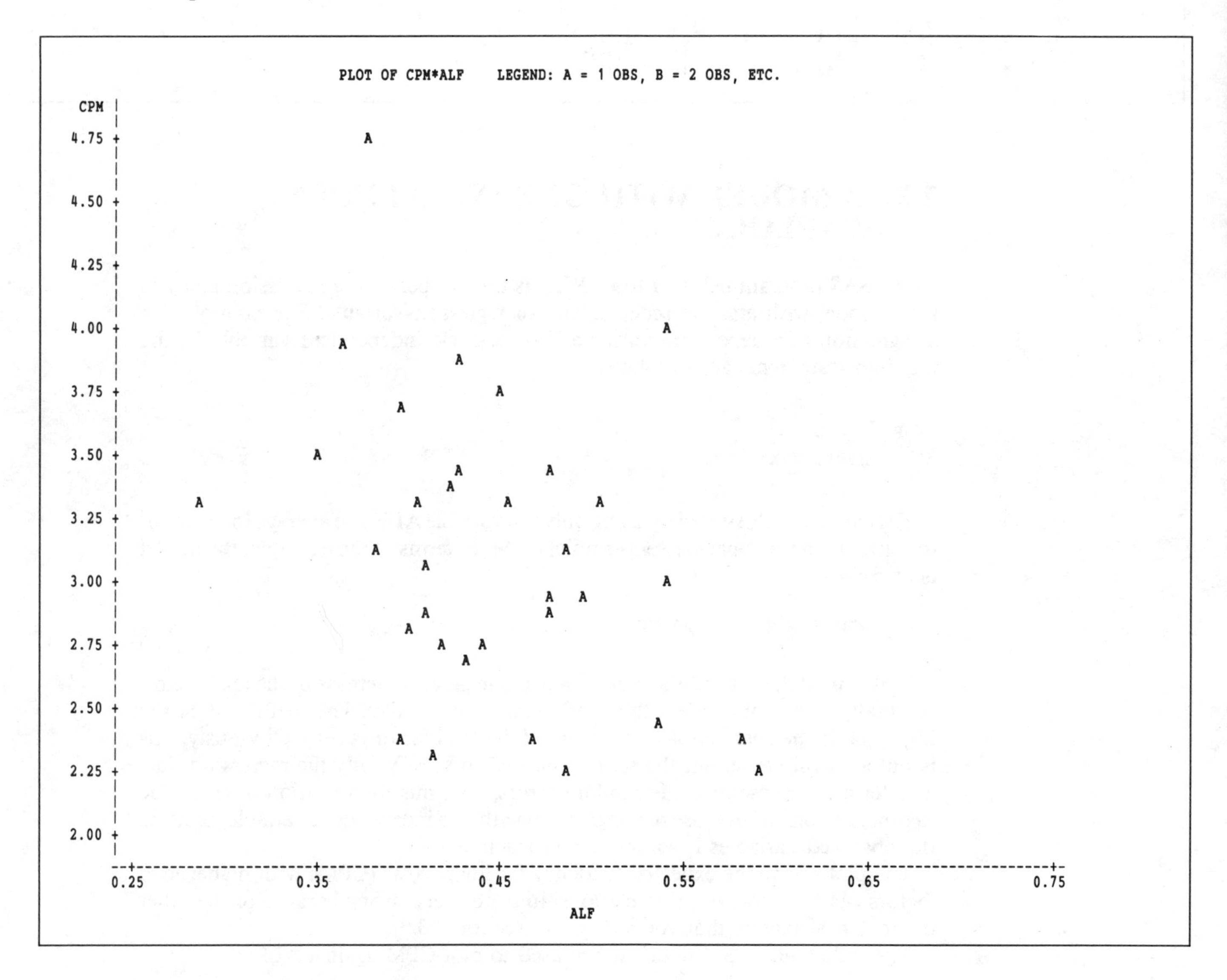

Output 2.2 Prediction Plot of CPM and ALF

The PROC statement invokes the REG procedure, and DATA=AIR tells the SAS System to apply PROC REG to the SAS data set AIR. The SAS System assigns the called procedure to the most recently created data set if the DATA= option is not indicated in the PROC statement. In the remainder of this book, the data set name will not be indicated, assuming that the appropriate data set was the last created.

The MODEL statement contains the equation CPM=ALF, which corresponds to the desired model. The left side specifies the dependent variable (CPM), and the right side specifies the independent variable (in this case ALF). The intercept term is not specified because PROC REG automatically assumes that it is to be estimated unless the NOINT option is used (see section **2.4.5**). Also, no term corresponding to the error term ε is indicated in the MODEL statement. PROC REG will produce ordinary least-squares estimates of the parameters. These estimates are optimal if the errors are independent and have equal variances. Methods for checking these assumptions and some suggestions for alternate methodologies are presented in Chapter 3.

The output from PROC REG is shown in **Output 2.3**. The circled numbers have been added to the output to key the descriptions that follow.

Output 2.3 Using PROC REG with DATA=AIR

```
❶
DEP VARIABLE: CPM
                                ANALYSIS OF VARIANCE
                              ❸
                       ❷      SUM OF      ❹ MEAN          ❺
         SOURCE        DF     SQUARES       SQUARE       F VALUE       PROB>F

         MODEL          1   1.52682210   1.52682210        5.034       0.0321
         ERROR         31   9.40185087   0.30328551
         C TOTAL       32  10.92867297
                                              ❼
            ❻ ROOT MSE     0.5507136     R-SQUARE         0.1397
              DEP MEAN     3.105697      ADJ R-SQ         0.1120
              C.V.         17.73237

                                PARAMETER ESTIMATES

                      ❾ PARAMETER    ❿ STANDARD   ⓫ T FOR H0:
❽ VARIABLE  DF          ESTIMATE          ERROR    PARAMETER=0      PROB > |T|

  INTERCEP   1        4.56054725     0.65545893          6.958          0.0001
  ALF        1       -3.26354830     1.45452657         -2.244          0.0321
```

1. The name of the dependent variable (CPM) identifies the analysis. This is especially useful if the MODEL statement specifies more than one independent variable or if there are multiple MODEL statements.
2. This column gives the degrees of freedom associated with the sums of squares discussed in 3, below.
3. The regression sum of squares (called MODEL SS; see section **1.1.2**) is 1.5268, and the residual sum of squares (called ERROR SS; see section **1.1.2**) is 9.4019. The sum of these two is the TOTAL SS of 10.9287, the corrected total. This illustrates the basic identity in regression analysis that TOTAL SS=MODEL SS+ERROR SS. This implies that variation among the observed values of the dependent variable can be attributed to two sources: (1) the variation that is due to changes in the independent variable and (2) residual variance that is not due to changes in the independent variable. If the model is correctly specified, this variance is due to random variation.
4. The corresponding mean squares are the sums of squares divided by their respective degrees of freedom. The MEAN SQUARE for error (ERROR) is an unbiased estimate of σ^2, the variance of ε, if the model has been correctly specified.
5. The value of the *F* statistic (F VALUE), 5.034, is the ratio of the model mean square divided by the error mean square. It is used to test the hypothesis that all coefficients except the intercept are zero. In this case, the hypothesis is that $\beta_1=0$. The *p* value (PROB>F) of 0.0321 indicates that there is about a 0.032 chance of obtaining an *F* value this large or larger if, in fact, $\beta_1=0$. Thus, there is reasonable evidence to state that $\beta_1 \neq 0$.
6. ROOT MSE=0.5507 is the square root of the error mean square. ROOT MSE estimates the standard deviation of ε. DEP MEAN=3.106 is simply the mean of all observed values of the dependent variable CPM. C.V.=17.73 is the coefficient of variation expressed as a percentage. This measure of relative variation is the ratio of ROOT MSE to the DEP

MEAN, times 100. In this example, the error standard deviation is 17.73% of the overall mean value of CPM. The coefficient is sometimes used as a standard to gauge the relative magnitude of error variation.

7. R-SQUARE=0.1397 is the square of the multiple correlation coefficient. For a one-variable regression, R-SQUARE is equivalent to the square of the correlation between the dependent and independent variables. It is also the ratio of MODEL SS divided by the TOTAL SS and thereby represents the fraction of total variation in the values of CPM explained by, or due to, the linear relationship to ALF (see 3, above). ADJ R-SQUARE, discussed in section **2.3**, is an alternative to R-SQUARE.
8. The column labeled VARIABLE identifies the regression coefficients. The label INTERCEP identifies β_0; other coefficients are identified by their respective variable names. In this case, the coefficient for ALF is the only other coefficient.
9. The values in the column labeled PARAMETER ESTIMATE are the estimated coefficients. The estimate of the intercept, β_0, is 4.561, and the estimate of the coefficient β_1 is -3.264. These estimates give the fitted model

$$\widehat{\text{CPM}} = 4.561 - (3.264)(\text{ALF})$$

where the caret (^) over CPM indicates that CPM is an estimate.

This expression can be used to estimate the average cost-per-passenger mile for a given average load factor. Section **2.4.1** shows how PROC REG can provide such estimates and their standard errors. The coefficient -3.264 also shows that, on the average, cost is decreased by 3.264 cents for each percent increase in the average load factor.

10. The estimated standard errors of the coefficient estimates are given in the column headed STANDARD ERROR. The estimates 0.655 and 1.455 can be used to construct confidence intervals for these model parameters. For example, a 95% confidence interval for β_1 is as follows:

$$(-3.265 \pm (2.042)(1.454)) = (-0.294, \; -6.232)$$

where 2.042 is the 0.05 level two-tail value of the t distribution for 30 degrees of freedom (used to approximate that for the 31 degrees of freedom needed here). Thus, with 95% confidence, the true relationship of cost to load factor is between -0.294 and -6.232 cents per passenger mile. Options for additional outputs, as well as instructions for plotting predicted and residual values, are presented in section **2.7.1** and Chapter 3.

11. The t statistics for testing the null hypothesis that each coefficient is zero are given in the column labeled T FOR H0: PARAMETER=0. These are simply the parameter estimates divided by their standard errors. The next column headed by PROB $>|\text{T}|$ gives the estimated p values (see the appendix) for these statistics. Thus, if you reject H_0: (coefficient for ALF=0), there is a 0.0321 chance of erroneous rejection. Note that for the one-variable model this is the same as the p value for the F test. Because the intercept has no practical value in the example, the test for this coefficient is not useful.

2.3 A MODEL WITH SEVERAL INDEPENDENT VARIABLES

The same example is used throughout this chapter to illustrate the use of PROC REG with various options. In **Output 2.5** PROC REG is used in a model with several independent variables in order to establish the relationship of cost-per-passenger mile to all the factors in the data except for TYPE.

Examining relationships among individual pairs of variables is not often useful in establishing the basis for a multiple regression model. Nevertheless, by examining the pairwise correlations among the five variables in the model, you can contrast these results with those using all variables simultaneously in a regression model. The correlations are obtained with the following statements:

```
PROC CORR;
   VAR ALF UTL ASL SPA CPM;
```

Output from these statements appears in **Output 2.4**.*

Output 2.4 PROC CORR Using DATA=AIR

VARIABLE	N	MEAN	STD DEV	SUM	MINIMUM	MAXIMUM
ALF	33	0.4457879	0.0669313	14.71100	0.2870000	0.591000
UTL	33	8.7172727	1.4490261	287.67000	5.6500000	13.300000
ASL	33	1.4690606	0.6499781	48.47900	0.8210000	3.607000
SPA	33	0.1835788	0.0897004	6.05810	0.0943000	0.359700
CPM	33	3.1056970	0.5843980	102.48800	2.2580000	4.737000

PEARSON CORRELATION COEFFICIENTS / PROB > |R| UNDER H0:RHO=0 / N = 33

	ALF	UTL	ASL	SPA	CPM
ALF	1.00000	-0.20538	-0.08719	-0.49490	-0.37378
	0.0000	0.2515	0.6295	0.0034	0.0321
UTL	-0.20538	1.00000	0.62842	0.32442	-0.37197
	0.2515	0.0000	0.0001	0.0655	0.0330
ASL	-0.08719	0.62842	1.00000	0.60710	-0.35078
	0.6295	0.0001	0.0000	0.0002	0.0453
SPA	-0.49490	0.32442	0.60710	1.00000	-0.29758
	0.0034	0.0655	0.0002	0.0000	0.0926
CPM	-0.37378	-0.37197	-0.35078	-0.29758	1.00000
	0.0321	0.0330	0.0453	0.0926	0.0000

The upper portion of the output reports the mean, standard deviation, and other statistics for the specified variables. The correlation coefficients appear in the array in the lower portion of the output. Each row and column in the array corresponds to a variable in the VAR (or VARIABLES) list. Two numbers appear in a given row and column. The upper number is the estimated correlation coefficient between the row variable and the column variable, and the lower number is the significance probability for testing that the corresponding population correlation

* PROC CORR is a comprehensive SAS software procedure for computing various types of correlation and covariance statistics. In this example only the simplest options are used. For more details see the description of this procedure in the *SAS User's Guide: Basics.*

is zero. For example, the estimated correlation between ALF and SPA is −0.4949, and this correlation is significantly different from zero at the p=0.0034 level. In other words, this correlation indicates evidence of a negative relationship between the average size of planes and the load factor. The correlations between the dependent variable, CPM, and the four independent variables are particularly interesting. All correlations are negative, and all, except for that with SPA, are significantly different from zero (p<0.05). This suggests that three of these factors can be useful, each by itself, in estimating the cost of providing air service.*

The relationship of CPM to the four cost factor variables in the model is estimated as follows:

$$CPM = \beta_0 + (\beta_1)(ALF) + (\beta_2)(UTL) + (\beta_3)(ASL) + (\beta_4)(SPA) + \varepsilon \quad .$$

To fit this model, use the following statements:

```
PROC REG;
   MODEL CPM=ALF UTL ASL SPA;
```

The output appears in **Output 2.5.**

Output 2.5 PROC REG with CPM and Four Cost Variables

```
DEP VARIABLE: CPM
                              ANALYSIS OF VARIANCE

                              SUM OF            MEAN
         SOURCE     DF       SQUARES          SQUARE         F VALUE ❶       PROB>F

         MODEL       4    6.57115408      1.64278852          10.556         0.0001
         ERROR      28    4.35751889      0.15562567
         C TOTAL    32   10.92867297

            ROOT MSE      0.3944942       R-SQUARE ❷       0.6013
            DEP MEAN       3.105697       ADJ R-SQ         0.5443
            C.V.           12.70228

                              PARAMETER ESTIMATES

                       PARAMETER          STANDARD       T FOR HO:
      VARIABLE  DF     ESTIMATE ❸         ERROR ❹    PARAMETER=0 ❺     PROB > |T|

      INTERCEP   1     8.59552505       0.90277548          9.521         0.0001
      ALF        1    -7.21137325       1.32056295         -5.461         0.0001
      UTL        1    -0.21281572       0.06508642         -3.270         0.0029
      ASL        1     0.33276893       0.18133342          1.835         0.0771
      SPA        1    -4.95030137       1.21695241         -4.068         0.0004
```

The upper portion of the output, as in **Output 2.3**, contains the partitioning of the sums of squares and the corresponding mean squares. The circled numbers have been added to the output to key the descriptions that follow.

* Note that the correlation between CPM and ALF is the square root of the R-SQUARE value and that the p value is the same for the regression between these variables (**Output 2.3**). This is true because the test for a correlation is also the test for a linear relationship between the variables. A regression relationship is a special case of a linear relationship.

1. The F value of 10.556 is used to test the null hypothesis:

$$H_0\colon \beta_1 = \beta_2 = \beta_3 = \beta_4 = 0 \quad .$$

The associated p value (PROB>F) of 0.0001 indicates that some of these coefficients are not zero and that the null hypothesis should be rejected.

2. R-SQUARE=0.6013 indicates that a major portion of the variation of CPM is due to variation in the independent variables in the model. Note that this R-SQUARE value is much larger than that for ALF alone (**Output 2.3**).

   ADJ R-SQ=0.5443 is an alternative to R-SQUARE that is adjusted for the number of parameters in the model according to the following formula:

$$\text{ADJ R-SQ} = 1 - (1 - \text{R-SQUARE})((n - 1) / (n - m - 1))$$

where n is the number of observations in the data set and m is the number of regression parameters in the model excluding the intercept. This adjustment measures the reduction in the mean square due to the regression. This adjustment is used to overcome an objection that R-SQUARE is a poor measure of goodness of fit because it can be forced to one (suggesting a perfect fit) simply by adding superfluous variables to the model with no real improvement to the fit. ADJ R-SQ tends to stabilize to a certain value when an adequate set of variables is included in the model. When m/n is small, say less than 0.05, the adjustment almost vanishes. Also, ADJ R-SQ can have values less than zero. Although not agreed upon uniformly, some authors also claim that this adjustment permits the comparison of regression models based on different sets of data.

3. The parameter estimates yield the equation for the fitted model:

$$\widehat{\text{CPM}} = 8.5955 - (7.2114)(\text{ALF}) - (0.2128)(\text{UTL}) + (0.3328)(\text{ASL}) - (4.9503)(\text{SPA}) \quad .$$

Thus, for example, a one percent increase in the average load factor (ALF) is associated with a decreased cost of 7.21 cents per passenger mile if all other factors are held constant.*

4. The estimated standard errors of the parameter estimates are useful for constructing confidence intervals, as illustrated in section **2.2**.

5. The t statistics are used for testing hypotheses about the individual parameters. It is important that you clearly understand the interpretation of these tests. The results can be explained in terms of comparing the fit of complete and reduced models (section **1.1.2**). The complete model for all of these tests contains all the variables on the right side of the MODEL statement. The reduced model contains all these variables except the one being tested. Thus, the t statistic of -5.461 for testing the hypothesis that $\beta_1=0$ (there is no effect due to ALF) is actually testing whether the complete four-variable model fits the data better than the reduced model containing only UTL, ASL, and SPA. In other words, the reduced model tests whether there is variation in CPM due to ALF that is not due to UTL, ASL, and SPA. The significance probability for this

---

* This coefficient is quite different from that obtained in the one-variable regression (**Output 2.3**). This is why simple, two-variable plots or correlations are not often useful in trying to determine the effects of variables in a multiple regression model.

test is 0.0001. By contrast, the corresponding significance probability for that variable in the one-variable regression was only 0.0321. Note that for UTL and SPA the p value for the parameter in the multiple variable model is smaller than that for the corresponding correlation (**Output 2.4**), while for ASL the reverse is true. This type of phenomenon results from correlations among the independent variables. This is discussed in more detail in the presentation of the Type I and Type II sums of squares (section **2.4.2**) and also in Chapter 4.

2.4 VARIOUS MODEL OPTIONS

The following section presents a number of MODEL options that provide additional optional output and modify the regression model. Although many of these options provide useful results, not all are useful for all analyses and should therefore be requested only if needed.

2.4.1 Model Options P, CLM, and CLI

One of the most common objectives of regression analysis is to compute the predicted values:

$$\hat{y} = \hat{\beta}_{(0)} + \hat{\beta}_{(1)} + \hat{\beta}_{(1)}x_1 + \ldots + \hat{\beta}_{(m)}x_m$$

for some selected values $x_1, \ldots, x_m$. This can be done in several ways using PROC REG. The most direct way is to use MODEL option P, implementing the following statements:

```
PROC REG;
   MODEL CPM=ALF / P CLM;
   ID ALF;
```

Note that one-variable regression is used for this section. The results of this option, as discussed here, are readily applicable to regressions with several independent variables. The purpose of the additional MODEL option CLM, as well as the ID statement, will be discussed later in this section. Specification of the P option in the MODEL statement causes the REG procedure to compute the $\hat{y}$ values corresponding to each observation in the data set. The results of these computations are printed following the basic PROC REG output. This additional output is presented in **Output 2.6**.

Output 2.6 Computing Predicted Values

OBS	❶ ID	❷ ACTUAL	❸ PREDICT VALUE	STD ERR PREDICT	LOWER95% MEAN	UPPER95% MEAN	❹ RESIDUAL
1	0.287	3.3060	3.6239	0.2501	3.1139	4.1339	-0.3179
2	0.349	3.5270	3.4216	0.1703	3.0742	3.7689	0.1054
3	0.362	3.9590	3.3791	0.1551	3.0629	3.6954	0.5799
4	0.378	4.7370	3.3269	0.1375	3.0465	3.6074	1.4101
5	0.381	3.0960	3.3171	0.1344	3.0430	3.5913	-0.2211

(continued on next page)

(continued from previous page)

6	0.394	3.6890	3.2747	0.1219	3.0261	3.5234	0.4143
7	0.397	2.3570	3.2649	0.1193	3.0217	3.5082	-0.9079
8	0.4	2.8330	3.2551	0.1167	3.0171	3.4932	-0.4221
9	0.405	3.3130	3.2388	0.1127	3.0089	3.4687	0.0742
10	0.409	3.0440	3.2258	0.1098	3.0018	3.4497	-0.1818
11	0.41	2.8460	3.2225	0.1091	3.0000	3.4450	-0.3765
12	0.412	2.3410	3.2160	0.1077	2.9963	3.4357	-0.8750
13	0.417	2.7800	3.1996	0.1046	2.9863	3.4130	-0.4196
14	0.422	3.3920	3.1833	0.1019	2.9755	3.3912	0.2087
15	0.425	3.8560	3.1735	0.1005	2.9685	3.3786	0.6825
16	0.426	3.4620	3.1703	0.1001	2.9661	3.3744	0.2917
17	0.434	2.7110	3.1442	0.0974	2.9455	3.3428	-0.4332
18	0.439	2.7430	3.1278	0.0964	2.9313	3.3244	-0.3848
19	0.452	3.7600	3.0854	0.0963	2.8890	3.2818	0.6746
20	0.455	3.3110	3.0756	0.0968	2.8782	3.2731	0.2354
21	0.466	2.4040	3.0397	0.1003	2.8352	3.2442	-0.6357
22	0.476	2.9620	3.0071	0.1055	2.7920	3.2222	-0.0451
23	0.476	3.4370	3.0071	0.1055	2.7920	3.2222	0.4299
24	0.478	2.9060	3.0006	0.1067	2.7829	3.2182	-0.0946
25	0.486	3.1400	2.9745	0.1123	2.7454	3.2035	0.1655
26	0.488	2.2750	2.9679	0.1138	2.7358	3.2001	-0.6929
27	0.495	2.9540	2.9451	0.1196	2.7011	3.1891	.0089092
28	0.504	3.3060	2.9157	0.1279	2.6549	3.1766	0.3903
29	0.535	2.4250	2.8145	0.1613	2.4855	3.1436	-0.3895
30	0.539	2.9710	2.8015	0.1660	2.4628	3.1402	0.1695
31	0.541	4.0240	2.7950	0.1684	2.4514	3.1385	1.2290
32	0.582	2.3630	2.6612	0.2201	2.2123	3.1101	-0.2982
33	0.591	2.2580	2.6318	0.2320	2.1587	3.1049	-0.3738
34	0.6	.	2.6024	0.2439	2.1049	3.0999	.

```
SUM OF RESIDUALS                6.92779E-14
SUM OF SQUARED RESIDUALS           9.401851
PREDICTED RESID SS (PRESS)         10.68407
```

The circled numbers have been added to the output to key the descriptions that follow.

1. The results from the ID statement identify each observation using the values of the indicated variable. In this case it is the independent variable ALF. Although one of the independent variables in the model is most frequently used as an ID variable, another variable, including character variables, can be used.
2. The P option provides the ACTUAL (observed) values of the dependent variable, CPM (compare with **Output 2.3**).
3. The column labeled PREDICT VALUE gives the set of values predicted with the estimated regression equation.
4. The set of residual values

   $$\text{RESIDUAL} = \text{ACTUAL} - \text{PREDICT}$$

   is also provided with the P option. It is sometimes useful to scan this column for relatively large values (see Chapter 3).

One of the handy features of PROC REG (as well as PROC GLM and PROC RSREG) is the ability to compute the $\hat{y}$ values for observations in the data set even if the observed value of the dependent variable is missing. Of course, complete data must be available for all the independent variables in the model. One additional observation that has the value 0.60 for ALF has been added to the data set AIR to illustrate this feature. All other values are missing. This observation cannot be used in the estimation of the model parameters, but it can be used to predict the cost-per-passenger mile for an average load factor using this one-variable model. The resulting cost is computed as follows:

$$4.561 - (3.264)(0.60) = 2.602 \quad .$$

This cost is the last item given in the column labeled PREDICT. The residual is, of course, not available. Additional examples of this feature are presented in section **2.7.1**. Chapter 5 illustrates how this feature can be used to construct plots of predicted values for an array of points that are not necessarily a part of the original data set.

The MODEL option CLM computes the 95% upper and lower confidence limits for the expected value of the dependent variable (mean) for each observation. These limits require the standard errors of the predicted values that you see listed in the output under the circled number 4. These limits are computed according to the following formula:

$$\text{STD ERR PREDICT} = ((\mathbf{x}'(\mathbf{X}'\mathbf{X})^{-1}\mathbf{x})(\text{MSE}))^{.5}$$

where $\mathbf{x}$ is a row vector of $\mathbf{X}$ corresponding to any observation and MSE is the error or residual mean square. This equation is equal to the square root of the variance of the subpopulation mean given in section **1.1.3**. The confidence limits are calculated with the following equation:

$$y_1 \pm t(\text{STD ERR PREDICT})$$

where t is the 0.05 level tabulated t value with degrees of freedom equal to those of the error mean square. For example, if you notice observation 1 on the output, ALF=3.306, the predicted value is 3.624 (under PREDICT VALUE), and the 95% confidence limits are 3.114 (LOWER 95% MEAN) to 4.134 (UPPER 95% MEAN).

Note that the standard error of the predicted value (STD ERR PREDICT) is also provided for the observation with the missing value of the dependent variable since the predicted value does not depend on the (actual) value of the dependent variable.

An additional MODEL option CLI (not used in this example) causes computation of the 95% prediction interval for a single observation. The formula for this is similar to that for the confidence interval for the mean. The difference is that the variance of the predicted value is larger than that for the interval of the mean by the value of the residual mean square. The results of the CLI option are not printed here; however, a plot of prediction intervals is presented in section **2.7.1**.

The interpretation of the CLM and CLI statistics is sometimes confused. Specifically, CLM yields a confidence interval for the subpopulation mean, and CLI yields a prediction interval for a single unit to be drawn at random from the population. The CLI limits will always be wider than the CLM limits because the CLM limits need to accommodate only the variability in the estimated mean, whereas the CLI limits must additionally accommodate the variability in the future single value of y. This is true even though the same predicted value ($\hat{y}$) is used both as the estimate of the subpopulation mean as well as the predictor of the future value.

To draw an analogy, suppose you walk into a roomful of people and are challenged to

1. estimate the average weight of all the people in the room
2. predict the weight of one particular person to be chosen at random.

You take a random sample and obtain a mean of 150 pounds. You would estimate the average weight to be 150 pounds, but you would also predict the weight of the mystery person to be 150 pounds; however, you would have more confidence in your estimate of the average weight of all the people than you would have in your prediction of the weight of the mystery person.

The following guidelines are offered to assist in deciding whether to use CLM or CLI:

1. Use CLM if you want the limits to show the region that should contain the population regression curve.
2. Use CLI if you want the limits to show the region that should contain (almost all of) the population of all possible observations.

Note: the confidence coefficients for CLM and CLI are valid on a single-point basis. Bands that are valid simultaneously for all points require computations not directly available in the SAS System.

Three statistics found at the bottom of **Output 2.6** are sometimes useful. These are the sum of residuals, the sum of squares of the actual residuals, and the PRESS statistic. The sum of residuals should be approximately zero, and the sum of squares of the actual residuals should equal the error sum of squares in the partitioning at the top of the PROC REG output. If different results are obtained, there is reason to question the results. Most likely, there has been excessive round off error (see section **2.4.5**). The PRESS statistic is the sum of squared residuals where each $\hat{y}$ is computed from a regression omitting that observation (see Chapter 3).

2.4.2 SS1 and SS2: Two Types of Sums of Squares

PROC REG can compute two types of sums of squares associated with the estimated coefficients in the model. These are referred to as Type I and Type II sums of squares. These are computed by specifying either the SS1 or the SS2 MODEL options. The following statements produce **Output 2.7** (the partitioning of the sums of squares and listing of parameter estimates are not affected by these options and are not reproduced):

```
PROC REG;
   MODEL CPM=ALF UTL ASL SPA / SS1 SS2;
```

Output 2.7 MODEL Options SS1 and SS2

DEP VARIABLE: CPM

VARIABLE	DF	TYPE I SS	TYPE II SS
INTERCEP	1	318.29667	14.10803605
ALF	1	1.52682210	4.64086546
UTL	1	2.29756722	1.66382597
ASL	1	0.17164558	0.52409629
SPA	1	2.57511918	2.57511918

The requested sums of squares are printed as additional columns in the section PARAMETER ESTIMATES and are labeled TYPE I SS and TYPE II SS. The interpretation of these sums of squares is based on the material in section **1.1.2**. In particular, the following concepts are useful in understanding the different types of sums of squares:

- partitioning of sums of squares
- complete and reduced models
- reduction notation.

The Type I sums of squares are commonly called sequential sums of squares. They represent a partitioning of the Model sums of squares into component sums of squares due to each variable as it is added sequentially to the model in the order prescribed by the MODEL statement. The TYPE I SS for the INTERCEP is simply $(\Sigma y)^2/n$, commonly called the correction for the mean. The TYPE I SS for ALF (1.5268) is the MODEL SS for a regression equation containing only ALF. This is easily verified in **Output 2.3**. The TYPE I SS for UTL (2.2976) is the increase in the MODEL SS due to adding UTL to a model already containing ALF. Because MODEL SS+ERROR SS=TOTAL SS, 2.2976 is also the decrease in the ERROR SS due to the addition of UTL to the model containing ALF. Equivalent interpretations hold for the TYPE I SS for ASL and SPA. In general terms, the Type I sum of squares for any particular variable is the increase in Model sum of squares due to adding that variable to a model that already contains all the variables preceding that variable in the MODEL statement. Note that the sum of the TYPE I SS is the overall MODEL SS:

$$6.5711 = 1.5268 + 2.2976 + 0.1716 + 2.5751 \quad .$$

This sum demonstrates the sequential partitioning of the Model sum of squares into Type I components corresponding to the variables as they are added to the model in the order given.

The Type II sums of squares are commonly called partial sums of squares. For a given variable, the Type II sum of squares is equivalent to the Type I sum of squares for that variable if that variable was the last in the MODEL list. This is easily verified by the fact that both types of sums of squares for SPA, which is the last variable in the list, are equal. In other words, the Type II sum of squares for a variable is the increase in Model sum of squares due to adding that variable to the model already containing all the other variables in the MODEL list. Therefore, the Type II sums of squares do not depend on the order in which the independent variables are listed in the MODEL statement. Furthermore, Type II sums of squares do not yield a partitioning of the Model sum of squares unless the independent variables are mutually uncorrelated. In reduction notation (see section **1.1.2**), the two types of sums of squares are as follows:

PARAMETER	TYPE I (Sequential)	TYPE II (Partial)
ALF	R(ALF | INTERCEPT)	R(ALF | INTERCEPT,UTL,ASL,SPA)
UTL	R(UTL | INTERCEPT,ALF)	R(UTL | INTERCEPT,ALF,ASL,SPA)
ALS	R(ASL | INTERCEPT,ALF,UTL)	R(ASL | INTERCEPT,ALF,UTL,SPA)
SPA	R(SPA | INTERCEPT,ALF,UTL,ASL)	R(SPA | INTERCEPT,ALF,UTL,ASL)

The reduction notation provides a convenient device to determine the complete and reduced models that are compared if the corresponding one degree of freedom sum of squares is used as the numerator for an *F* test. *F* tests derived by dividing the Type II sums of squares by the error mean square are equivalent to the *t* tests for the parameters provided in the computer output. In fact, the Type II *F* statistic is equal to the square of the *t* statistic.

Tests derived from the Type I sums of squares are convenient to use in model building in cases where there is a predetermined order for selecting variables. Not only does each Type I sum of squares provide a test for the improvement in the fit of the model when the corresponding term is added to the model, but the additivity of the sums of squares can be used to assess the significance of a model containing, for example, the first $k<m$ terms. Associated with the Type I sums of squares is the MODEL option SEQB. This option prints the coefficients of the successive models fitted by adding terms in the order given in the MODEL statement. Type I sums of squares and the SEQB option are very useful in fitting polynomial models. Section **5.2** presents this use of these options.

PROC REG does not provide the F tests associated with the Type I and Type II sums of squares. This causes no difficulty for the Type II sums of squares because these are the squares of the corresponding t statistics. Therefore, F ratios corresponding to the Type I sums of squares must be calculated manually.*

2.4.3 Standardized Coefficients: the STB Option

The STB option can be used in the MODEL statement to produce the set of standardized regression coefficients. **Output 2.8** contains the PARAMETERS ESTIMATE portion of the output from the following statements:

```
PROC REG;
   MODEL CPM=ALF UTL ASL SPA / STB;
```

Output 2.8 Producing Model Parameters

PARAMETER ESTIMATES

VARIABLE	DF	PARAMETER ESTIMATE	STANDARD ERROR	T FOR H0: PARAMETER=0	PROB > |T|	STANDARDIZED ESTIMATE
INTERCEP	1	8.59552505	0.90277548	9.521	0.0001	0
ALF	1	-7.21137325	1.32056295	-5.461	0.0001	-0.82592086
UTL	1	-0.21281572	0.06508642	-3.270	0.0029	-0.52768069
ASL	1	0.33276893	0.18133342	1.835	0.0771	0.37011165
SPA	1	-4.95030137	1.21695241	-4.068	0.0004	-0.75983144

The coefficients labeled STANDARDIZED ESTIMATE are the estimates obtained if all variables in the model were standardized to zero mean and unit variance prior to performing the regression computations. Therefore, the magnitudes of these coefficients are not affected by the scales of measurement of the various model variables. Each coefficient indicates the number of standard deviation changes in the dependent variable associated with a standard deviation change in the independent variable if all other variables are held constant.

* If it is essential to have these F ratios provided in a computer output, they are available with PROC GLM. PROC GLM, which is primarily designed for less than full-rank models, is useful for performing analysis of variance and covariance by regression methods (Freund and Littell 1986). For full-rank models of the type discussed in this book, the Types II, III, and IV sums of squares produced by PROC GLM provide the same results.

2.4.4 Printing Matrices: the XPX and I Options

The XPX and I options are available to specify the printing of the matrices used in the regression computations (sections **1.1.1**, **1.1.2**, and **1.1.4**). The XPX option, standing for "**X** prime **X**," prints the following matrix:

$$\begin{bmatrix} X'X & X'Y \\ Y'X & Y'Y \end{bmatrix} .$$

In other words, this matrix is the matrix of sums of squares and cross products of all the variables in the MODEL statement. The I option, standing for "Inverse," prints the following matrix:

$$\begin{bmatrix} (X'X)^{-1} & \hat{\beta} \\ (\hat{\beta}') & \text{Error SS} \end{bmatrix} .$$

The output (not including the usual regression output) from the following statements is shown in **Output 2.9**:

```
PROC REG;
   MODEL CPM=ALF UTL ASL SPA / XPX I;
```

Output 2.9 PROC REG Using the XPX and I Options

MODEL CROSSPRODUCTS X'X X'Y Y'Y

X'X	INTERCEP	ALF	UTL
INTERCEP	33	14.711	287.67
ALF	14.711	6.701339	127.6024
UTL	287.67	127.6024	2574.887
ASL	48.479	21.48998	441.5446
SPA	6.0581	2.605547	54.15947
CPM	102.488	45.22007	883.3362

X'X	ASL	SPA	CPM
INTERCEP	48.479	6.0581	102.488
ALF	21.48998	2.605547	45.22007
UTL	441.5446	54.15947	883.3362
ASL	84.73768	10.03238	146.2974
SPA	10.03238	1.369616	18.31544
CPM	146.2974	18.31544	329.2253

X'X INVERSE, B, SSE

INVERSE	INTERCEP	ALF	UTL
INTERCEP	5.236948	-6.6065	-0.264323
ALF	-6.6065	11.20565	0.1644021
UTL	-0.264323	0.1644021	0.02722071
ASL	0.5549365	-0.640166	-0.0471844
SPA	-4.20857	6.092528	0.1256214
CPM	8.595525	-7.21137	-0.212816

(continued on next page)

(continued from previous page)

INVERSE	ASL	SPA	CPM
INTERCEP	0.5549365	-4.20857	8.595525
ALF	-0.640166	6.092528	-7.21137
UTL	-0.0471844	0.1256214	-0.212816
ASL	0.2112878	-0.918588	0.3327689
SPA	-0.918588	9.516252	-4.9503
CPM	0.3327689	-4.9503	4.357519

The rows and columns of the matrices are identified by the names of the variables in the model. The variable INTERCEP corresponds to a dummy variable whose value is unity for all observations. The INTERCEP variable is used to estimate the intercept coefficient (β_0). Thus, in the **X′X** matrix the first row (or column) consists of the sample size and the sums of the variables. All other elements are sums of squares and cross products of the variables. The portion of the inverse corresponding to INTERCEP and the independent variables is the inverse of **X′X**. The row (or column) corresponding to the dependent variable contains the coefficient estimates (compare to **Output 2.4**). The last element, which has both row and column indicating the name of the dependent variable, contains the error sum of squares. If there are several dependent variables, then the appropriate number of rows and columns corresponding to the dependent variables containing the coefficient estimates and error sums of squares will be present.

Additional options, COVB and CORRB, print matrices. COVB prints the matrix of variances and covariances. CORRB prints the matrix of correlations of the estimated coefficients. The option ALL prints all of the statistics corresponding to the various model options discussed, plus some descriptive statistics of the model variables.

2.4.5 Regression through the Origin: the NOINT Option

The NOINT option can be used in the MODEL statement to force the regression response to pass through the origin. In other words, the estimated value of the dependent variable is zero when all independent variables have the value zero. An example of this requirement occurs in some growth models where the response (for instance, weight) must be zero at the beginning, that is, when time is zero. In many applications this requirement is not reasonable, especially when this condition does not or cannot actually occur. If this option is implemented in such situations, the results of the regression analysis are often grossly misleading. Moreover, even when the conditions implied by the NOINT option are reasonable, the results of the analysis have some features that may mislead the unwary practitioner. The NOINT option can be illustrated by implementing it on the airline data. This example clearly demonstrates the uselessness of the results when this option is not justified. This happens because zero values of the variables in the model cannot occur. The regression is implemented by using the following statements:

```
PROC REG;
   MODEL CPM=ALF UTL ASL SPA / NOINT P;
```

Note that in addition to the NOINT option, the P option is specified to obtain the predicted values. The results are shown in **Output 2.10**.

Output 2.10 PROC REG with the NOINT Option

```
DEP VARIABLE: CPM
                              ANALYSIS OF VARIANCE

                          SUM OF           MEAN
        SOURCE     DF    SQUARES          SQUARE         F VALUE       PROB>F

        MODEL       4   310.75979   77.68994727         122.011        0.0001
        ERROR      29 18.46555493    0.63674327
        U TOTAL    33   329.22534

           ROOT MSE     0.7979619       R-SQUARE          0.9439
           DEP MEAN      3.105697       ADJ R-SQ          0.9362
           C.V.          25.69349
NOTE: NO INTERCEPT TERM IS USED. R-SQUARE IS REDEFINED.

                              PARAMETER ESTIMATES

                       PARAMETER         STANDARD       T FOR H0:
  VARIABLE  DF          ESTIMATE            ERROR     PARAMETER=0      PROB > |T|

  ALF        1        3.63202333       1.35217330           2.686          0.0118
  UTL        1        0.22102456       0.09400915           2.351          0.0257
  ASL        1       -0.57806125       0.31159738          -1.855          0.0738
  SPA        1        1.95732639       1.97632390           0.990          0.3302

                                          PREDICT
                   OBS       ACTUAL         VALUE     RESIDUAL

                     1       3.3060        2.6366       0.6694
                     2       3.5270        2.7570       0.7700
                     3       3.9590        3.0898       0.8692
                     4       4.7370        2.3996       2.3374
                     5       3.0960        3.2487      -0.1527
                     6       3.6890        2.9286       0.7604
                     7       2.3570        2.9600      -0.6030
                     8       2.8330        3.1537      -0.3207
                     9       3.3130        3.3594      -0.0464
                    10       3.0440        3.2583      -0.2143
                    11       2.8460        3.4019      -0.5559
                    12       2.3410        2.8401      -0.4991
                    13       2.7800        2.7355       0.0445
                    14       3.3920        3.0495       0.3425
                    15       3.8560        2.5304       1.3256
                    16       3.4620        3.0421       0.4199
                    17       2.7110        2.9350      -0.2240
                    18       2.7430        2.8566      -0.1136
                    19       3.7600        2.8861       0.8739
                    20       3.3110        2.8790       0.4320
                    21       2.4040        3.4065      -1.0025
                    22       2.9620        3.3846      -0.4226
                    23       3.4370        2.7413       0.6957
                    24       2.9060        3.0813      -0.1753
                    25       3.1400        3.2980      -0.1580
                    26       2.2750        3.1124      -0.8374
                    27       2.9540        3.3919      -0.4379
                    28       3.3060        3.3730      -0.0670
                    29       2.4250        3.6856      -1.2606
                    30       2.9710        3.1119      -0.1409
                    31       4.0240        3.0684       0.9556
                    32       2.3630        3.1237      -0.7607
                    33       2.2580        3.1204      -0.8624

SUM OF RESIDUALS                   1.641323
SUM OF SQUARED RESIDUALS           18.46555
```

The overall partitioning of the SUM OF SQUARES and the R-SQUARE statistic certainly suggests a well-fitting model. In fact, both the MODEL F and the R-SQUARE statistics are much larger than those for the model with the intercept (**Output 2.5**). However, closer examination of the results shows that the error sum of squares for the no-intercept model is larger than the total sum of squares for the model with the intercept. This apparent contradiction arises from the fact that in the no-intercept model the total sum of squares is the uncorrected or uncentered sum of squares, Σy^2, for the dependent variable. In contrast, the total sum of squares for the model with the intercept is the corrected or centered sum of squares, $\Sigma(y-\bar{y})^2$. Although the error sum of squares is much larger for the no-intercept model, the difference between the total and error sum of squares is much larger for the no-intercept model, giving a larger value to the R-SQUARE statistic. These differences are recognized by the output, which designates the total sum of squares as U TOTAL for the no-intercept model and C TOTAL for the intercept model. Furthermore, both total and error degrees of freedom are larger by one in the no-intercept model because one less parameter has been estimated. Finally, the warning message

```
NOTE: NO INTERCEPT TERM IS USED. R-SQUARE IS REDEFINED.
```

is intended to alert the reader to the special interpretation of this statistic.

Looking further, you can also see that the coefficient estimates for the no-intercept model bear no resemblance to those of the intercept model. Finally, the statistics that follow the listing of predicted and residual values show that the sum of residuals is not zero. This is in contrast to models with intercepts where the sum of residuals is, by definition, equal to zero. It is a feature of no-intercept models that the sum of residuals is not zero, although, if the true intercept is near zero, the sum of residuals may be quite close to zero.

A simple example can illustrate the fact that the NOINT option may provide misleading results, even in cases where the true intercept is near zero. The model $y=x$ generates eight data points with a normally distributed error having zero mean and unit variance. The data are given in **Output 2.11**.

Output 2.11 Sample Data to Illustrate the NOINT Option

OBS	X	Y
1	1	-.35
2	2	2.79
3	3	1.81
4	4	2.00
5	5	3.88
6	6	6.79
7	7	7.67
8	8	6.79

Then, regression models with and without intercept are implemented using the sample data. The SAS statements are as follows;

```
PROC REG;
   MODEL Y=X;
   MODEL Y=X / NOINT;
```

The results of these analyses are shown in **Output 2.12** and **Output 2.13**.

Output 2.12 PROC REG Results without the NOINT Option

```
DEP VARIABLE: Y
                              ANALYSIS OF VARIANCE

                              SUM OF          MEAN
         SOURCE     DF       SQUARES        SQUARE       F VALUE          PROB>F

         MODEL       1   49.50857143   49.50857143        34.232          0.0011
         ERROR       6    8.67757857    1.44626310
         C TOTAL     7   58.18615000

             ROOT MSE       1.202607      R-SQUARE        0.8509
             DEP MEAN         3.9225      ADJ R-SQ        0.8260
             C.V.           30.65919

                              PARAMETER ESTIMATES

                     PARAMETER          STANDARD      T FOR H0:
  VARIABLE  DF        ESTIMATE             ERROR    PARAMETER=0        PROB > |T|

  INTERCEP   1     -0.96321429        0.93706366         -1.028            0.3436
  X          1      1.08571429        0.18556626          5.851            0.0011


                                               PREDICT
                         OBS      ACTUAL         VALUE      RESIDUAL

                           1     -0.3500        0.1225       -0.4725
                           2      2.7900        1.2082        1.5818
                           3      1.8100        2.2939       -0.4839
                           4      2.0000        3.3796       -1.3796
                           5      3.8800        4.4654       -0.5854
                           6      6.7900        5.5511        1.2389
                           7      7.6700        6.6368        1.0332
                           8      6.7900        7.7225       -0.9325

SUM OF RESIDUALS               1.77636E-15
SUM OF SQUARED RESIDUALS          8.677579
```

Output 2.13 PROC REG Results with the NOINT Option

```
DEP VARIABLE: Y
                              ANALYSIS OF VARIANCE

                              SUM OF          MEAN
         SOURCE     DF       SQUARES        SQUARE       F VALUE          PROB>F

         MODEL       1     171.06851     171.06851       117.335          0.0001
         ERROR       7   10.20568971    1.45795567
         U TOTAL     8     181.27420

             ROOT MSE       1.207458      R-SQUARE        0.9437
             DEP MEAN         3.9225      ADJ R-SQ        0.9357
             C.V.           30.78288
NOTE: NO INTERCEPT TERM IS USED. R-SQUARE IS REDEFINED.
```

(continued on next page)

(continued from previous page)

PARAMETER ESTIMATES

VARIABLE	DF	PARAMETER ESTIMATE	STANDARD ERROR	T FOR H0: PARAMETER=0	PROB > |T|
X	1	0.91573529	0.08453899	10.832	0.0001

OBS	ACTUAL	PREDICT VALUE	RESIDUAL
1	-0.3500	0.9157	-1.2657
2	2.7900	1.8315	0.9585
3	1.8100	2.7472	-0.9372
4	2.0000	3.6629	-1.6629
5	3.8800	4.5787	-0.6987
6	6.7900	5.4944	1.2956
7	7.6700	6.4101	1.2599
8	6.7900	7.3259	-0.5359

```
SUM OF RESIDUALS               -1.58647
SUM OF SQUARED RESIDUALS       10.20569
```

You can immediately see that both the MODEL F and the R-SQUARE values are much larger for the no-intercept model, whereas the residual mean squares (ERROR MEAN SQUARE) are almost identical. Actually, both models have estimated very similar regression lines because the estimated intercept (-0.963) is so close to zero that a hypothesis test of a zero intercept cannot be rejected ($p=0.3436$). However, the residual mean square of the no-intercept model is still somewhat larger, and the sum of residuals (obtained with the P option but not shown here) is zero for the intercept model and -1.586 for the no-intercept model.

2.5 RESTRICTED LEAST SQUARES

The no-intercept model is a special case of restricted least-squares models where the coefficient estimates are subject to one or more linear restrictions. PROC REG allows restricted least-squares estimation with the use of RESTRICT statements. The RESTRICT statement follows a MODEL statement and has the following general form:

RESTRICT *equation1,equation2,equation3,*
equation4,
.
.
.
equationk ;

where each equation is a linear combination of model parameters set equal to a constant. Thus, for example, the no-intercept model can be implemented by adding the following statement and not using the no-intercept MODEL option:

```
RESTRICT INTERCEPT=0;
```

The restrict option is used with the example given in **Output 2.11**; the results are shown in **Output 2.14**. The restriction is, of course, the one implied by the NOINT option, so you can compare the results.

Output 2.14 Illustrating the Restrict Statement

```
DEP VARIABLE: Y
                              ANALYSIS OF VARIANCE

                          SUM OF          MEAN
        SOURCE    DF     SQUARES        SQUARE       F VALUE       PROB>F

        MODEL      0  47.98046029                      0.000       1.0000
        ERROR      7  10.20568971   1.45795567
        C TOTAL    7  58.18615000

            ROOT MSE     1.207458      R-SQUARE       0.8246
            DEP MEAN       3.9225      ADJ R-SQ       0.8246
            C.V.         30.78288

                              PARAMETER ESTIMATES

                      PARAMETER        STANDARD     T FOR H0:
  VARIABLE  DF         ESTIMATE           ERROR   PARAMETER=0        PROB > |T|

  INTERCEP   1     -2.22045E-16     1.00581E-08        -0.000            1.0000
  X          1       0.91573529      0.08453899        10.832            0.0001
  RESTRICT  -1      -1.58647059      1.54962536        -1.024            0.3400
```

Note that with this method the total sum of squares is centered; hence, the R-SQUARE value has the same connotation as in intercept models. However, the MODEL DF is zero since you are comparing the model sum of squares of two one-parameter models; hence, the F statistic is meaningless. In fact, this F ratio can be negative as, for example, in the AIR data.

The coefficient estimate and standard error for X are the same as for the NOINT option. The last line (RESTRICT) gives the test for the reduction in MODEL SS due to the elimination of the intercept. Note that the p value for this test is the same as the test for the intercept (except for a small round off error) shown in **Output 2.12**.

2.6 TESTS FOR SUBSETS AND LINEAR FUNCTIONS OF PARAMETERS

The t tests on the parameters in the basic PROC REG output (**Output 2.3**) provide tests of hypotheses that individual regression coefficients are equal to zero. Section **2.4.2** indicated that the Type I sums of squares can be used to test that certain subsets of coefficients are zero. This section illustrates how the TEST statement can be used to test whether one or more linear functions of parameters are equal to specified constants. As a special case, the TEST statement tests whether one or more subsets of coefficients are equal to zero. The TEST statement must accompany a MODEL statement in PROC REG. Several TEST statements may follow any one MODEL statement. The general form of a TEST statement is as follows:

label: **TEST** *equation1*,
equation2,
.
.
.
equationk ;

The label is optional and serves only to identify specific tests in the output. The equations specify the linear functions to be tested. The tests can be interpreted in terms of comparing complete and reduced models in the manner described in previous sections. The complete model for all tests specified by a TEST statement is the model containing all variables on the right side of the MODEL statement. The reduced model is derived from the complete model by imposing the restrictions implied by the equations specified in the TEST statement. For this example the following statements are added to the MODEL statement in section **2.3**:

```
TEST1 : TEST SPA,ALF;
TEST2 : TEST UTL,ALF;
TEST3 : TEST UTL-ALF=0;
```

The first TEST statement (TEST1) tests the hypothesis that both coefficients for SPA and ALF are zero. This statement shows one feature of these statements: if the value of the function to be tested is zero, the equality need not be specified. In other words, this statement can also be written:

```
TEST1 : SPA=0, ALF=0;
```

This test compares the full model to one containing only UTL and ASL. The second TEST statement (TEST2) tests the hypothesis that both coefficients for UTL and ALF are zero. The third TEST statement (TEST3) tests the hypothesis that the sum of coefficients for UTL and ALF is zero. This is equivalent to the test that the coefficient for UTL is equal to the coefficient for ALF. (This particular test has no practical interpretation in this example and is given here for illustration only.) The results of these TEST statements that appear right after the output of parameter estimates are given in **Output 2.15**.

Output 2.15 Testing Subsets

```
TEST: TEST1         NUMERATOR:      2.40585  DF:     2   F VALUE:     15.4592
                    DENOMINATOR: 0.155626  DF:    28   PROB >F :     0.0001

TEST: TEST2         NUMERATOR:      2.55124  DF:     2   F VALUE:     16.3935
                    DENOMINATOR: 0.155626  DF:    28   PROB >F :     0.0001

TEST: TEST3         NUMERATOR:      4.49189  DF:     1   F VALUE:     28.8634
                    DENOMINATOR: 0.155626  DF:    28   PROB >F :     0.0001
```

For each TEST statement indicated, a sum of squares is computed with degrees of freedom equal to the number of equations in the TEST statement. From these quantities a mean square is computed that forms the numerator of an F statistic. The denominator of this statistic is the error mean square for the full model. The values of the two mean squares, the degrees of freedom, the F ratio, and its p value are printed on the output. (Note: if linear dependencies or inconsistencies exist among the equations of a TEST statement, PROC REG prints a message that the test failed, and no F ratio is computed.)

2.7 CREATING OUTPUT DATA SETS

It is often instructive to use the statistics generated by the P, R, CLM, and other MODEL options for additional analyses. The output data set abilities of PROC REG (as well as many other SAS procedures) can facilitate such analyses.

Occasionally, some results of regression analyses may be useful for further analyses. Data sets containing the results of the options mentioned above can also be provided by PROC REG (see section **2.7.2**).

2.7.1 The OUTPUT Statement

In PROC REG, you can use an OUTPUT statement following the MODEL statement to construct a new SAS data set. This data set contains all of the variables in the data set to which PROC REG was applied, plus other variables specified in the OUTPUT statement. The basic form of the OUTPUT statement to create a data set containing predicted and residual values is as follows:

OUTPUT OUT=*SASdataset* **P=***names* **R=***names*;

where *SASdataset* is the name chosen for the new data set, **P=***names* is a list of names chosen for the variables whose values are the predicted values, and **R=***names* is the list of names chosen for the variables whose values are residuals. The names in the **P=***names* and **R=***names* lists must correspond to the list of dependent variables in the MODEL statement. A wide variety of output statistics can be included in the OUTPUT data set. Some are produced in the example in this section, and others are given in Chapter 3. The *SAS User's Guide: Statistics* gives the complete set of statistics that can be included.

The OUTPUT statement can be illustrated by implementing the one-variable regression from section **2.2**. This produces a data set that includes predicted values and the 95% prediction intervals. Then, the results are plotted. The following SAS statements are used:

```
PROC REG;
   MODEL CPM=ALF;
   OUTPUT OUT=D P=PCPM R=RCPM U95=UP L95=DOWN;
PROC PLOT;
   PLOT CPM*ALF='.' PCPM*ALF='P' UP*ALF='U' DOWN*ALF='D' / OVERLAY;
```

The output from PROC REG is the same as in **Output 2.3**. The OUTPUT statement specifies the following:

OUT=D	specifies that the name of the new data set is D.
P=PCPM	specifies that the predicted values have the variable name PCPM.
R=RCPM	specifies that the residuals have the variable name RCPM.
U95=UP	specifies that the upper 95% prediction limit has the variable name UP.
L95=DOWN	specifies that the lower 95% prediction limit has the variable name DOWN.

Alternatively, you can compute the upper and lower 95% confidence intervals on the mean (see section **2.4.1**) using the U95M and L95M options.

As a result of the OUTPUT statement, the SAS log carries the additional notation:

```
NOTE: DATA SET WORK.D HAS 33 OBSERVATIONS AND 10 VARIABLES.
```

This notation shows that the procedure has created the new data set. The number of variables comprises the original set of six plus the four additional variables specified in the OUTPUT statement. Now PROC PLOT is used to plot the actual, predicted, and 95% prediction interval values; the residuals are not used in this plot (see section **2.7.2** and Chapter 3). The plot of CPM*ALF is shown in **Output 2.2.**

2.7.2 Options for Producing Other Results: OUTEST, OUTSSCP, and COVOUT Options

Results of the PROC REG calculations can be output to a SAS data set by using the following PROC options:

OUTEST=*SASdataset*
: produces a data set containing the parameter estimates and the residual standard deviation for each MODEL. The additional COVOUT option adds the matrix of variances and covariances of the coefficient estimates to the data set.

  This type of data set may be useful if a regression model has been applied to data sets corresponding to different populations or treatments. It is also useful to determine how the regression coefficients differ among the populations.

OUTSSCP=*SASdataset*
: produces a data set containing the sums of squares and cross products of all variables in the MODEL statements.

The OUTSSCP= data sets can be used directly as input to many other statistical procedures in the SAS System. The OUTSSCP= option can be especially useful when a large number of observations are explored in many different runs.

The printer plot does not perform the best job of displaying these statistics; however, if your site licenses SAS/GRAPH software, you can take advantage of its features to enhance your plots. SAS/GRAPH software allows more flexible spacing and interpolation options. The required steps (in addition to hardware specific instructions) follow:

```
PROC GPLOT DATA=D;
   PLOT CPM*ALF=1
        PCPM*ALF=2
        UP*ALF=3
        DOWN*ALF=4 / OVERLAY;
   SYMBOL1 V=STAR C=BLACK;
   SYMBOL2 V=P I=JOIN C=BLACK;
   SYMBOL3 V=U I=SPLINE C=BLACK;
   SYMBOL4 V=L I=SPLINE C=BLACK;
```

This graph is shown in **Output 2.16.**

Output 2.16 Regression Using SAS/GRAPH Software

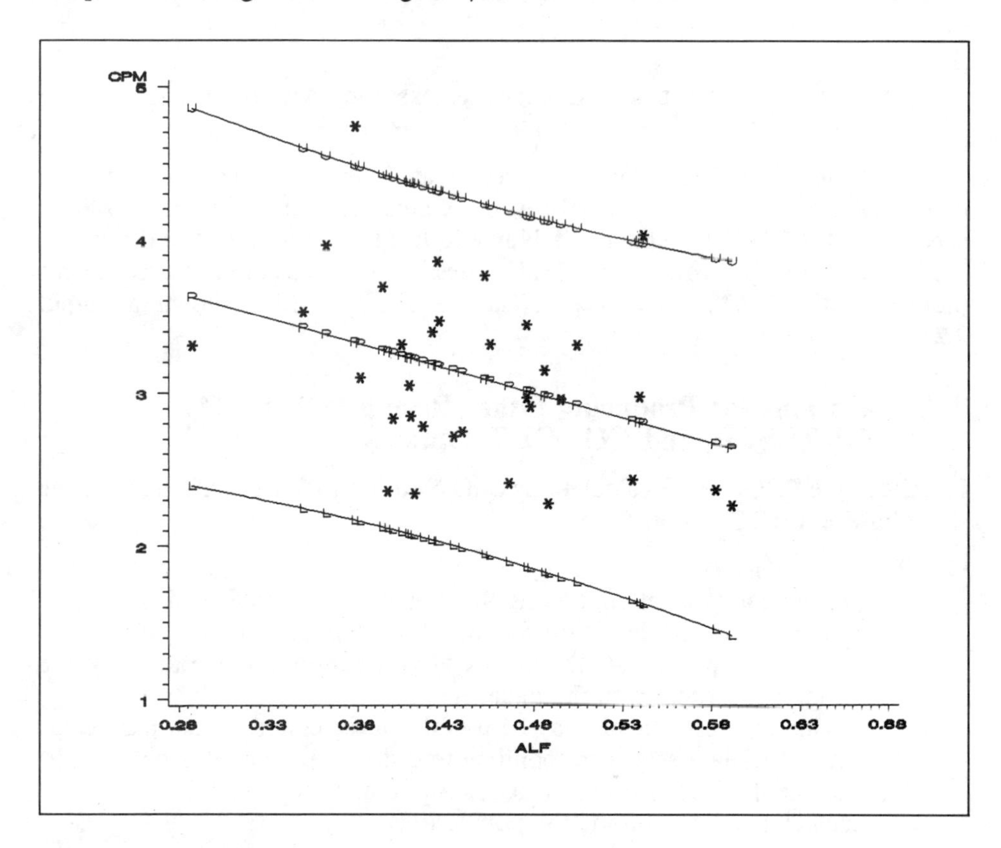

2.7.3 Predicting to a Different Set of Data

A regression equation estimated from one set of data can be used to predict values of the dependent variable for another set of similar data. This type of prediction has been used to settle charges of pay discrimination against minorities. An equation relating pay to factors such as education, performance, and tenure is estimated for nonminority employee data. This equation predicts what the pay rate should be for all employees in the absence of discrimination or other factors. If this equation predicts salaries substantially higher than actual salaries for minority employees, then there is cause to suspect pay discrimination.

The airline cost data can illustrate this type of analysis. The variable TYPE was created to divide the airlines into two groups: the short-haul lines with ASL<1200 miles and the long-haul lines with ASL≥1200 miles. (This is an arbitrary division used here only for illustration.) Comparing the cost equation for short-haul lines to the cost equation for the long-haul lines is one way to determine if there are differences in their cost structures.*

Define a new variable, SCPM, to give the cost-per-passenger mile of the short-haul lines. This variable is missing for the other airlines. The following statements are inserted before the CARDS statement:

```
IF TYPE=0 THEN SCPM=CPM;
ELSE SCPM=.;
```

* Another method for answering this question is given in section **6.6**.

Now invoke PROC REG:

```
PROC REG;
   MODEL SCPM=ALF UTL ASL SPA;
   OUTPUT OUT=E P=PSCPM R=RSCPM;
```

The regression estimates from this analysis are based only on the fourteen short-haul lines.

Output 2.17 Regression Analysis Using TYPE= to Create a Subset

```
DEP VARIABLE: SCPM
                                 ANALYSIS OF VARIANCE

                         SUM OF          MEAN
        SOURCE    DF    SQUARES         SQUARE        F VALUE       PROB>F

        MODEL      4   3.90857548   0.97714387        17.486        0.0003
        ERROR      9   0.50292195   0.05588022
        C TOTAL   13   4.41149743

            ROOT MSE     0.23639      R-SQUARE        0.8860
            DEP MEAN    3.335571      ADJ R-SQ        0.8353
            C.V.        7.086941

                                 PARAMETER ESTIMATES

                     PARAMETER         STANDARD        T FOR H0:
  VARIABLE  DF        ESTIMATE            ERROR      PARAMETER=0       PROB > |T|

  INTERCEP   1     10.70003668       1.11685114            9.581           0.0001
  ALF        1     -6.72039997       1.65218520           -4.068           0.0028
  UTL        1     -0.41388562       0.05584654           -7.411           0.0001
  ASL        1     -0.27370554       0.65235057           -0.420           0.6846
  SPA        1     -5.49132038       3.45924778           -1.587           0.1469
```

Note that the coefficients are somewhat different from those estimated using the entire data set (**Output 2.3**). The most striking difference is that the number of seats per plane is not statistically significant (p=0.1469). This is probably due to short-haul lines not having a wide assortment of airplane sizes. Obviously, data set E does not have residuals for the long-haul lines; therefore, these residuals are computed in another step:

```
DATA F;
   SET E;
   RSCPM=CPM-PSCPM;
```

Next, the residuals are plotted against the predicted values using the value of TYPE as the plotting symbol. A reference line at the zero value of the residuals is also requested.

```
PROC PRINT;
PROC PLOT;
   PLOT RSCPM*PSCPM=TYPE / VREF=0;
```

The resulting plot is shown in **Output 2.18**.

Output 2.18 A Plot of Residual and Predicted Values

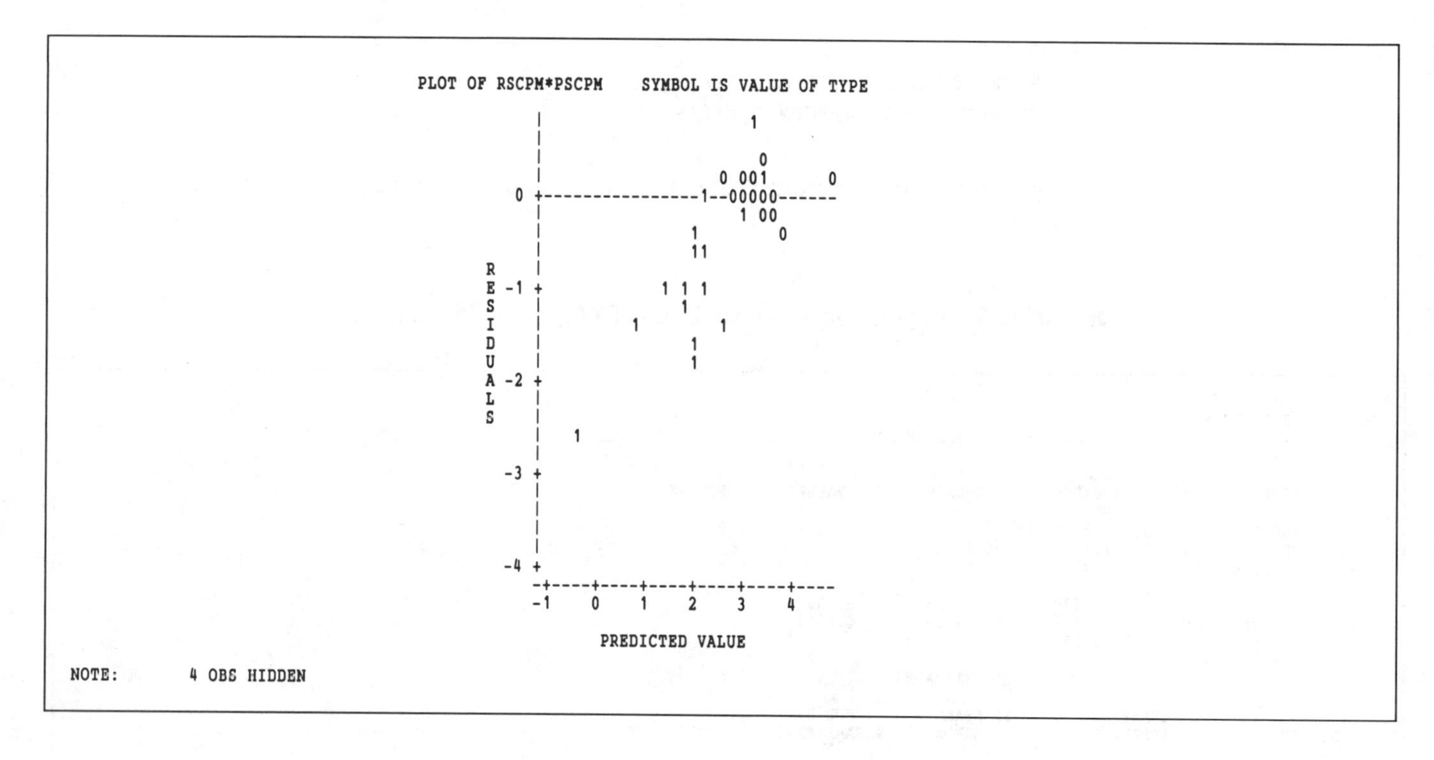

You can see that the cost equation of the short-haul lines does not predict well the costs of the long-haul lines. In fact, the long-haul lines appear to have substantially lower costs.

2.8 EXACT COLLINEARITY: LINEAR DEPENDENCY

Linear dependency occurs when exact linear relationships exist among the independent variables. More precisely, a linear dependency exists when one or more columns of the **X** matrix can be expressed as a linear combination of other columns. This means that the **X′X** matrix is singular and cannot be inverted in the usual sense to obtain parameter estimates.*

PROC REG is programmed to detect the existence of exact collinearity, and if it exists, PROC REG uses a generalized inverse (section **1.1.4**) to compute "parameter estimates." "Parameter estimates" is placed in quotation marks to emphasize that care must be exercised to determine exactly what parameters are being estimated. More technically, the generalized inverse approach yields one particular solution to the normal equations. The PROC REG computations with exact linear dependencies are illustrated here with an alternate model of the airline

* Exact collinearity should not be confused with multicollinearity, a term used to describe a high degree of multiple correlation among the independent variables. Multicollinearity can be described as almost exact collinearity. Thus, multicollinearity is a related topic, but in fact, it represents an entirely different condition requiring different methodology. Multicollinearity and associated methodology are discussed in Chapter 4.

cost data. First, you define a new variable in the DATA step:

```
PASS=ALF*SPA;
```

This is the average number of passengers per flight. Next use a model in which all variables have been converted to logarithms. (The logarithmic model is discussed in detail in section **6.2**). The DATA step accomplishes this as follows:

```
ARRAY A*ALF USL ASL SPA PASS CPM;
DO OVER A;
   A=LOG(A);
END;
```

Now implement PROC REG using all five variables:

```
PROC REG;
   MODEL CPM=ALF UTL ASL SPA PASS;
```

The output is shown in **Output 2.19**.

Output 2.19 All Variables Converted to Logarithms

```
DEP VARIABLE: CPM
                                 ANALYSIS OF VARIANCE

                            SUM OF          MEAN
        SOURCE     DF      SQUARES        SQUARE        F VALUE        PROB>F

        MODEL       4   0.70233524    0.17558381         12.364        0.0001
        ERROR      28   0.39763916    0.01420140
        C TOTAL    32   1.09997441

            ROOT MSE     0.1191696      R-SQUARE         0.6385
            DEP MEAN      1.116472      ADJ R-SQ         0.5869
            C.V.          10.67377

NOTE: MODEL IS NOT FULL RANK. LEAST SQUARES SOLUTIONS FOR THE
      PARAMETERS ARE NOT UNIQUE. SOME STATISTICS WILL BE
      MISLEADING. A REPORTED DF OF 0 OR B MEANS THAT THE
      ESTIMATE IS BIASED. THE FOLLOWING PARAMETERS HAVE BEEN
      SET TO 0, SINCE THE VARIABLES ARE A LINEAR COMBINATION
      OF OTHER VARIABLES AS SHOWN.
PASS    =+1*ALF      +SPA

                                 PARAMETER ESTIMATES

                          PARAMETER          STANDARD       T FOR H0:
  VARIABLE  DF             ESTIMATE             ERROR     PARAMETER=0        PROB > |T|

  INTERCEP   1           0.63878695        0.39010340           1.637            0.1127
  ALF        B          -1.03377453        0.18011628          -5.739            0.0001
  UTL        1          -0.46778757        0.15840127          -2.953            0.0063
  ASL        1           0.11080915        0.08472969           1.308            0.2016
  SPA        B          -0.33716783        0.08083128          -4.171            0.0003
  PASS       0                    0                 .               .                 .
```

The NOTE in the model summary indicates the existence of collinearity. The NOTE is followed by the equation describing the collinearity:

```
PASS =+1*ALF +SPA
```

This collinearity arises because PASS=ALF*SPA, log(PASS)=log(ALF) +log(SPA), thus constituting an exact collinearity.

The "parameter estimates" that are printed are equivalent to those obtained if PASS was not included in the MODEL statement. In general, the "parameter estimates" are those that would be obtained if any variable were deleted from the MODEL statement, providing that variable is a linear function of variables preceding it in the MODEL statement.

These deleted variables are indicated with zeros under DF and PARAMETER ESTIMATE headings. Other variables involved in the linear dependencies are indicated with a B, standing for bias, under the DF heading. These estimates are, in fact, unbiased estimates of the parameters of the model that does not include the (deleted) variables (those indicated with a 0) but are biased estimates for other models.

The bias can be readily seen by implementing PROC REG with the independent variables listed in a different order in the MODEL statement. For example, if you specify

```
MODEL CPM=PASS ALF UTL ASL SPA;
```

then SPA has DF and PARAMETER ESTIMATE=0 because SPA=PASS−ALF, whereas the coefficient estimates for SPA and ALF have different values that are denoted as biased. Note further that any coefficients not involved (PASS in this example) are not biased, and the coefficients retain the same estimated value under any ordering of variables in the MODEL statement.

Of course, finding a linear dependency does not directly lead to a course of action. It is obvious that one variable in a set of variables involved in the dependency should be deleted, but the decision regarding the variable to delete is up to you.

Because of round off errors, determining when exact linear dependency occurs is somewhat arbitrary. In PROC REG the matrix inversion procedure computes successive tolerance values. A tolerance of zero signifies an exact collinearity, but because round off errors essentially preclude the computing of exact zero values, a tolerance of 10^{-8} is normally used as sufficiently close to zero to indicate exact collinearity. A different criterion can be invoked by including the PROC REG statement option

```
SINGULAR=t;
```

where *t* specifies the minimum tolerance for which the matrix is declared nonsingular.

2.9 MULTIPLE MODEL STATEMENTS

A single invocation of PROC REG can contain several MODEL statements. Although the use of multiple models in one PROC REG implementation appears to offer savings in computer resources, this is not always the case. Furthermore, the use of multiple MODEL statements may incur some unwanted side effects.

When PROC REG encounters multiple MODEL statements, it computes a single **X′X** matrix containing all variables from all models. From this matrix, PROC REG picks the subset of the overall **X′X** matrix corresponding to the list of dependent and independent variables for each specified model. PROC REG does this for every model and computes all statistics. If the several MODEL statements have a large number of overlapping variables, the same elements of the overall **X′X** matrix are used several times, thus saving computer resources. However, if there are few or no variables common to the several models, a large number of elements that have been computed for the overall **X′X** matrix are not used, and computer resources have been wasted. When computing the overall **X′X** matrix, PROC REG must, by necessity, eliminate from the computations any observation having a missing value for any variable to be included in the matrix. In some cases this procedure excludes some observations that have information for one or more of the individual models. Thus, if there are missing values in the data, multiple MODEL statements are not recommended.

3. Observations

3.1 INTRODUCTION

In the linear model

$$y = X\beta + \varepsilon$$

the vector ε is assumed to be a random variable specified to be independently and normally distributed with mean zero and variance σ^2. This variable is called the random error, and the individual elements are called random errors. They are said to represent "natural variation" but may also measure the cumulative effects of factors not specified by the other model parameters. Violations of this assumption can occur in many ways. The most frequent occurrences can be categorized as follows:

- The data may contain *outliers*, or unusual observations, that do not belong to the population specified by the model.
- A *specification error* occurs when the specified model does not contain all of the necessary parameters. This can result when some important independent variables have not been included in the model or if only linear terms have been specified and the true relationships are nonlinear.
- The distribution of the variable may be distinctly nonnormal; the distribution may be severely skewed or fat-tailed.
- The random variable may exhibit *heteroscedasticity*; in other words, the variances are not the same for the entire population.
- The individual errors may be correlated. This is a phenomenon usually found in time series data, but it is not restricted to such situations.

Often violations of the assumptions underlying the random errors are not so severe as to invalidate the analysis, but this is not always guaranteed. For this reason, it is useful to look for evidence regarding possible violations of these assumptions.

In performing a regression analysis, individual elements of the population of random errors are estimated by the sample of residuals

$$\hat{\varepsilon} = y - X\hat{\beta} \quad .$$

If the assumptions of the model apply to the data, they should behave as a sample from the population specified for the εs.

This chapter presents some tools available in PROC REG to detect some of these violations. This chapter also presents alternative methodologies that you can use if assumptions fail. This coverage is not exhaustive, especially with respect to alternative methodologies (see Belsley et al. 1980).

3.2 OUTLIER DETECTION

Observations that do not appear to fit the model, often called *outliers*, can be quite troublesome because they can influence parameter estimates and make the resulting analysis less useful. For this reason, you must carefully examine your statistical analysis to determine if any observations exist that may cause misleading results. In this context it should be noted that an outlier may be "unusual" with respect to the values of the independent variables or the dependent variable. Each of these conditions can create different types of misleading results. You should be especially careful when you use regression analyses because the lack of structure of the independent variables makes detection and identification of outliers more difficult.

The following example consists of data collected to determine the effect of certain variables on the efficiency of irrigation. The dependent variable is the percent of water percolation (PERC), and the independent variables are listed below:

RATIO is the ratio between irrigation time and advance time.

INFT is the exponent of time in the infiltration equation, a calculated value.

LOST is the percentage of water lost to deep percolation.

ADVT is the exponent of time in the water advance equation, a calculated value.

The data are presented in **Output 3.1**.

Output 3.1 Irrigation Data

RATIO	INFT	LOST	ADVT	PERC
0.77	0.427	29.10	0.5820	37.75
0.97	0.427	29.10	0.6980	34.83
1.15	0.309	20.90	0.5190	33.75
1.27	0.427	29.10	0.8800	30.26
1.27	0.309	20.90	0.6850	30.60
1.51	0.309	20.90	0.8000	29.05
1.67	0.343	21.27	0.5640	25.76
1.87	0.309	20.90	0.8360	26.26
2.25	0.309	20.90	0.8410	25.75
2.32	0.397	26.53	0.5960	17.16
2.34	0.343	21.27	0.7500	14.73
2.39	0.427	29.10	0.6000	19.04
2.71	0.397	25.45	0.7470	13.16
2.82	0.427	29.10	0.7300	18.03
3.17	0.397	26.53	0.4360	14.46
3.35	0.397	26.53	0.7600	12.96
3.64	0.343	21.27	0.6991	16.80
3.69	0.387	26.53	0.6000	13.72
3.73	0.397	25.45	0.7010	14.56
3.78	0.387	25.45	0.6800	12.71
4.97	0.397	26.53	0.7720	9.06
6.86	0.397	26.53	0.8400	9.52

First implement the regression analysis using the following SAS statements:

```
PROC REG;
   MODEL PERC=RATIO INFT LOST ADVT;
```

The output is shown in **Output 3.2**.

Output 3.2 Regression Analysis Using Irrigation Data

```
DEP VARIABLE: PERC
                              ANALYSIS OF VARIANCE

                             SUM OF          MEAN
          SOURCE      DF    SQUARES        SQUARE        F VALUE       PROB>F

          MODEL        4  1300.19826     325.04956        16.234       0.0001
          ERROR       17   340.38669   20.02274630
          C TOTAL     21  1640.58495

             ROOT MSE      4.474678      R-SQUARE         0.7925
             DEP MEAN      20.90545      ADJ R-SQ         0.7437
             C.V.          21.40436

                              PARAMETER ESTIMATES

                        PARAMETER        STANDARD       T FOR H0:
    VARIABLE  DF        ESTIMATE           ERROR      PARAMETER=0        PROB > |T|

    INTERCEP   1     40.55825454     11.34544931            3.575            0.0023
    RATIO      1     -4.82052470      0.74296236           -6.488            0.0001
    INFT       1      -209.31100       100.93107           -2.074            0.0536
    LOST       1      2.75482588      1.35638415            2.031            0.0582
    ADVT       1      4.28506704      8.82267965            0.486            0.6334
```

The regression is certainly statistically significant. The variable RATIO appears to be very significant; INFT and LOST contribute marginally, while ADVT appears useless.

3.2.1 Residuals and Studentized Residuals

Examining the estimated residuals, that is, the differences between the actual values of the dependent variable and those predicted by the regression equation, is a traditional method for detecting outliers (as well as specification error, see section **3.3**). It is important to note that a statistical analysis using the SAS System, or any other software system, is used only to help detect outliers. You must decide what to do if outliers are found.

As shown in Chapter 2, these residuals can be printed as part of the PROC REG output by using the P (or PREDICTED) option in the MODEL statement, or these residuals can be output to a data set that can be used to plot the residuals. You can examine these results, especially the plots, for residuals having large magnitudes that may correspond to outliers.

A difficulty with estimated residuals is that they are not all estimated with the same precision. The standard errors of estimated residuals can be computed, and when the residuals are divided by these standard errors, standardized or *studentized* residuals are obtained. If the assumptions about the random error hold, these studentized residuals can be considered to be a sample from Student's t distribution. Values from the t distribution exceeding 2.5 in absolute value are relatively rare, except when the error degrees of freedom are small. Thus, these studentized residuals provide a convenient vehicle for identifying unusually large residuals. Such studentized residuals, together with other related statistics, are available with the R (or RESIDUAL) option in the MODEL statement. Also, you can output the statistics provided by this option to a data set for plotting (or other purposes). Modify the SAS statements as follows:

```
PROC REG;
   MODEL PERC=RATIO INFT LOST ADVT / R;
   OUTPUT OUT=A P=PPERC R=RPERC STUDENT=STUDENT;
   ID RATIO;
```

The OUTPUT statement (see section **2.7**) causes the creation of a data set called A. In addition to the variables in the original data set, data set A includes three new variables:

PPERC	are the predicted values.
RPERC	are the residual values.
STUDENT	are the studentized residuals.

The ID statement causes the values of the specified variable (RATIO in this case) to be printed in all outputs. This helps to identify the observations that may be associated with suspected outliers. The additional output from PROC REG is given in **Output 3.3**.

Output 3.3 Residual Statistics

```
                             PREDICT   STD ERR              STD ERR   STUDENT                       COOK'S
OBS      ID     ACTUAL        VALUE    PREDICT   RESIDUAL   RESIDUAL  RESIDUAL    -2-1-0 1 2             D

  1    0.77    37.7500      30.1300     2.2207     7.6200     3.8848    1.9615   |     |***   |      0.251
  2    0.97    34.8300      29.6630     2.1160     5.1670     3.9428    1.3105   |     |**    |      0.099
  3    1.15    33.7500      30.1374     2.5954     3.6126     3.6451    0.9911   |     |*     |      0.100
  4    1.27    30.2600      28.9967     2.7905     1.2633     3.4980    0.3612   |     |      |      0.017
  5    1.27    30.6000      30.2702     1.9925     0.3298     4.0066    0.0823   |     |      |      0.000
  6    1.51    29.0500      29.6061     2.0938    -0.5561     3.9546   -0.1406   |     |      |      0.001
  7    1.67    25.7600      21.7262     2.5367     4.0338     3.6862    1.0943   |     |**    |      0.113
  8    1.87    26.2600      28.0250     2.1821    -1.7650     3.9065   -0.4518   |     |      |      0.013
  9    2.25    25.7500      26.2146     2.1878    -0.4646     3.9034   -0.1190   |     |      |      0.001
 10    2.32    17.1600      21.9176     1.3398    -4.7576     4.2694   -1.1144   |   **|      |      0.024
 11    2.34    14.7300      19.2935     2.3199    -4.5635     3.8263   -1.1927   |   **|      |      0.105
 12    2.39    19.0400      22.3979     1.7577    -3.3579     4.1150   -0.8160   |    *|      |      0.024
 13    2.71    13.1600      17.7094     1.9169    -4.5494     4.0433   -1.1252   |   **|      |      0.057
 14    2.82    18.0300      20.8821     1.6410    -2.8521     4.1629   -0.6851   |    *|      |      0.015
 15    3.17    14.4600      17.1345     2.5627    -2.6745     3.6681   -0.7291   |    *|      |      0.052
 16    3.35    12.9600      17.6552     1.2432    -4.6952     4.2985   -1.0923   |   **|      |      0.020
 17    3.64    16.8000      12.8087     2.1876     3.9913     3.9035    1.0225   |     |**    |      0.066
 18    3.69    13.7200      17.4237     2.0243    -3.7037     3.9906   -0.9281   |    *|      |      0.044
 19    3.73    14.5600      12.5954     1.7484     1.9646     4.1189    0.4770   |     |      |      0.008
 20    3.78    12.7100      14.3575     1.2711    -1.6475     4.2903   -0.3840   |     |      |      0.003
 21    4.97     9.0600       9.8974     1.9094    -0.8374     4.0468   -0.2069   |     |      |      0.002
 22    6.86     9.5200       1.0780     3.1458     8.4420     3.1823    2.6528   |     |***** |      1.375

SUM OF RESIDUALS             5.65326E-13
SUM OF SQUARED RESIDUALS        340.3867
```

The various statistics are identified by column headings and include the ID variable, the previously defined ACTUAL, predicted (PREDICT VALUE), RESIDUAL, and studentized residual values (STUDENT RESIDUAL), as well as the standard errors of the predicted (STD ERR PREDICT) and residual values (STD ERR RESIDUAL).*

A plot of the studentized residuals follows these statistics. Each * represents a magnitude of 0.5. The final column contains the Cook's D statistic that is discussed later in this chapter. The residuals as well as the studentized residuals are plotted against the predicted values to show more clearly what these statistics can reveal.

```
PROC PLOT DATA=A;
   PLOT (RPERC STUDENT)*PPERC / HPOS=30 VPOS=25 VREF=0;
```

The plots are shown in **Output 3.4** and **3.5**.

Output 3.4 Using PROC PLOT RPERC*PPERC

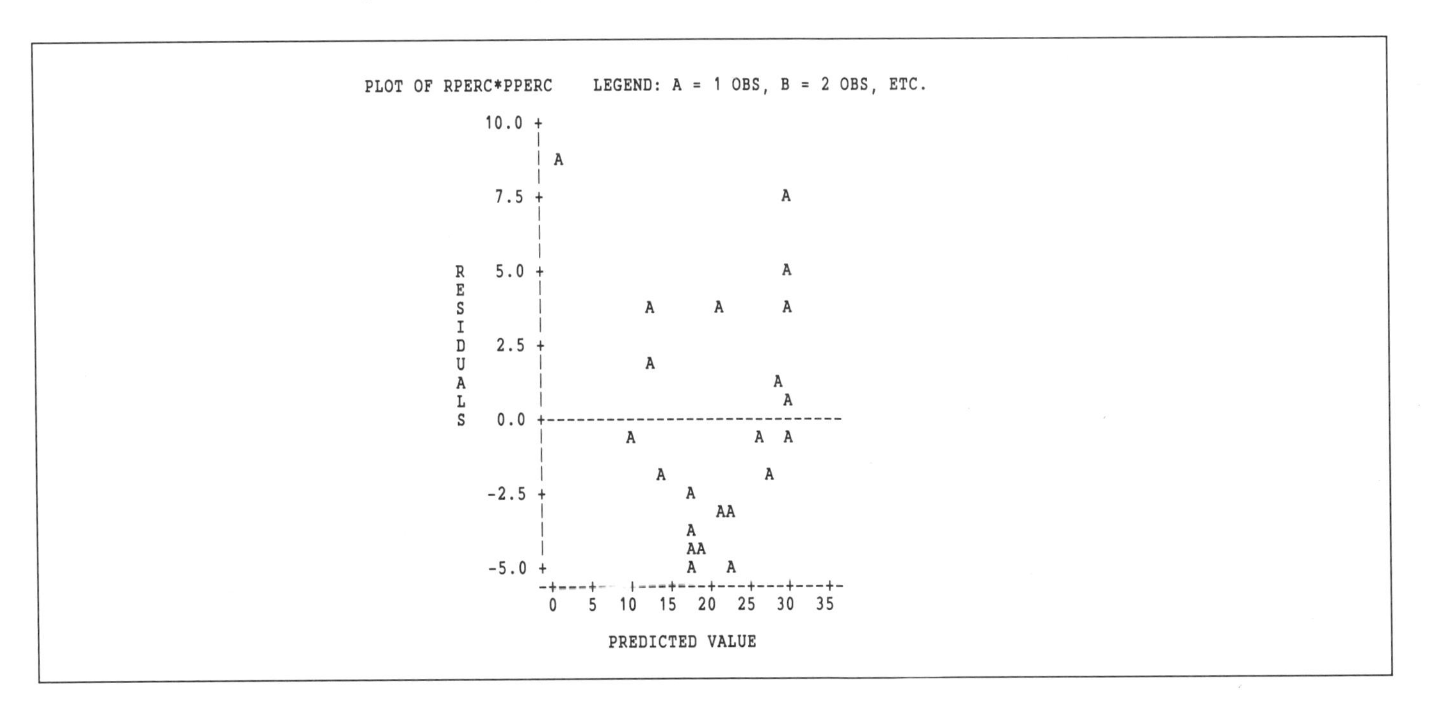

* These standard errors can be included in the output data set by using the keywords STP and STR together with the variable names desired for these variables.

Output 3.5 Using PROC PLOT STUDENT*PPERC

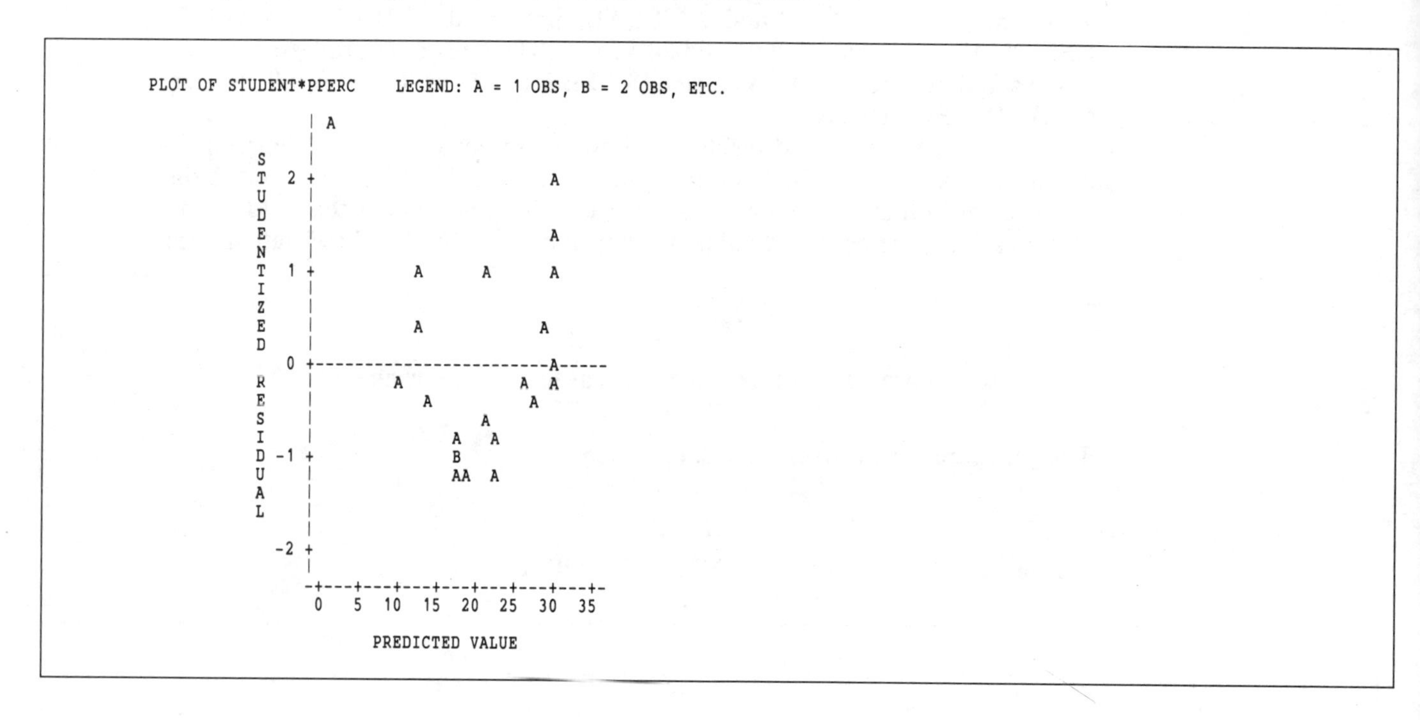

You can see that the studentized residuals have almost the same pattern as the actual residuals. This occurs in this example because the standard errors of the residuals do not vary much. This is readily confirmed by the values of these standard errors (column labeled STD ERR RESIDUAL in **Output 3.3**). However, if there are no outliers, the distribution of the studentized residuals should conform to Student's *t* distribution with 17 degrees of freedom (see the ERROR DF, **Output 3.2**). Therefore, the value of the last observation (3.182) is somewhat suspicious. This particular observation is also suspect because the predicted value is far lower than that for any other observation. Another interesting feature is that the next to largest residual is at the other extreme of the range of predicted values. Thus, there is some evidence that the last observation is an outlier; however, the evidence is not very convincing.

3.2.2 Influence Statistics

One reason that you cannot always detect outliers when you examine the residuals is that the least-squares estimation procedure tends to pull the estimated regression response towards outlying observations. Estimated residuals for such observations may not be especially large, thus hindering the search for outliers. You may be able to overcome this difficulty by asking what would happen to the outlier detection statistics if the observation in question were not used in the estimation of the regression equation used to calculate the statistics. Statistics of this type are used to determine the potential influence of a particular observation; hence, they are often called *influence statistics*. PROC REG provides a set of such statistics with the INFLUENCE option in the MODEL statement. Modify the PROC

REG statements as follows:

```
PROC REG;
   MODEL PERC=RATIO INFT LOST ADVT / R INFLUENCE;
   OUTPUT OUT=A P=PPERC R=RPERC RSTUDENT=RSTUDENT DFFITS=DFFITS;
   ID RATIO;
```

The INFLUENCE option adds information to the output already produced in **Output 3.3** to give the results shown in **Output 3.6**.

Output 3.6 Using the INFLUENCE Option

OBS	ID	RESIDUAL	RSTUDENT	HAT DIAG H	COV RATIO	DFFITS	INTERCEP DFBETAS	RATIO DFBETAS	INFT DFBETAS	LOST DFBETAS	ADVT DFBETAS
1	0.77	7.6200	2.1635	0.2463	0.4980	1.2367	-0.1974	-0.7056	0.0216	0.1508	-0.2904
2	0.97	5.1670	1.3409	0.2236	1.0240	0.7196	-0.3368	-0.4715	0.0543	0.0601	0.1575
3	1.15	3.6126	0.9906	0.3364	1.5154	0.7053	0.6391	-0.0133	-0.3320	0.2470	-0.4492
4	1.27	1.2633	0.3517	0.3889	2.1321	0.2806	-0.2001	-0.1593	0.0355	0.0027	0.2021
5	1.27	0.3298	0.0799	0.1983	1.6856	0.0397	0.0282	-0.0068	-0.0201	0.0147	-0.0042
6	1.51	-0.5561	-0.1365	0.2190	1.7236	-0.0723	-0.0273	0.0149	0.0310	-0.0226	-0.0261
7	1.67	4.0338	1.1011	0.3214	1.3849	0.7578	0.2193	-0.2316	0.4951	-0.5599	-0.2543
8	1.87	-1.7650	-0.4410	0.2378	1.6725	-0.2463	-0.0726	0.0278	0.1055	-0.0778	-0.1133
9	2.25	-0.4646	-0.1155	0.2390	1.7720	-0.0647	-0.0201	-0.0007	0.0302	-0.0226	-0.0288
10	2.32	-4.7576	-1.1229	0.0897	1.0180	-0.3524	-0.0564	0.0360	-0.0107	-0.0135	0.1944
11	2.34	-4.5635	-1.2087	0.2688	1.1963	-0.7328	0.0715	0.2234	-0.5573	0.6160	-0.2083
12	2.39	-3.3579	-0.8076	0.1543	1.3112	-0.3450	0.0360	0.0206	0.0666	-0.1172	0.1331
13	2.71	-4.5494	-1.1347	0.1835	1.1263	-0.5379	0.2860	0.1739	-0.4471	0.4231	-0.2001
14	2.82	-2.8521	-0.6740	0.1345	1.3601	-0.2657	0.1395	0.0139	0.0413	-0.0868	-0.0624
15	3.17	-2.6745	-0.7187	0.3280	1.7190	-0.5021	-0.2513	-0.1693	0.0798	-0.0828	0.4538
16	3.35	-4.6952	-1.0989	0.0772	1.0198	-0.3178	0.1426	-0.0637	-0.0064	-0.0202	-0.1297
17	3.64	3.9913	1.0239	0.2390	1.2955	0.5738	0.0756	0.0987	0.3579	-0.4225	-0.0112
18	3.69	-3.7037	-0.9241	0.2047	1.3127	-0.4688	-0.2066	-0.2804	0.3022	-0.3067	0.2638
19	3.73	1.9646	0.4659	0.1527	1.4940	0.1977	-0.0657	0.0277	0.1437	-0.1396	0.0123
20	3.78	-1.6475	-0.3742	0.0807	1.4102	-0.1109	-0.0015	-0.0584	-0.0190	0.0188	0.0222
21	4.97	-0.8374	-0.2010	0.1821	1.6348	-0.0948	0.0175	-0.0708	0.0180	-0.0205	-0.0148
22	6.86	8.4420	3.3619	0.4942	0.1850	3.3234	-0.4485	2.8230	-0.7752	0.7800	0.5600

The first two columns repeat the ID variable and the actual residuals. The remainder of the statistics are described below:

RSTUDENT residuals
: are studentized residuals. The studentized residual is the residual divided by its standard error which uses $s^2_{(i)}$, without the ith observation, not s^2 which is used for the STUDENT RESIDUAL.

HAT DIAG H statistics
: are the diagonal elements of the hat matrix, $\mathbf{X}(\mathbf{X}'\mathbf{X})^{-1}\mathbf{X}'$. These values, often denoted h_i, measure the influence (also called leverage) of each observation. Essentially, observations with large h_i have the potential of causing trouble if they are outliers. If m is the number of independent variables, the sum of h_i is $(m+1)$; hence, the average value is approximately $(m+1)/n$, where n is the number of observations in the data set. Values of this statistic exceeding $2(m+1)/n$ may be considered indicators of influential observations.

COV RATIO statistics
: measure the change in the determinant of the $\mathbf{X}'\mathbf{X}$ matrix caused by deleting the observation. Observations that cause large changes in this matrix may be considered outliers with respect to the pattern of independent variables. These observations may be influential

observations. The average value of this statistic is unity; deviations from unity exceeding $3(m+1)/n$ are considered large.

DFFITS statistics
: measure the difference in the estimated or predicted value for an observation using the equation estimated by all observations and the equation estimated from all other observations. The difference is standardized using the variance estimate ($s^2_{(i)}$) from the all other observations equation. Belsley et al. (1980, 28) suggest that DFFITS values exceeding $2\sqrt{(m+1)/n}$ are convenient criteria for identifying suspected outliers. These criteria account for the fact that magnitudes of the DFFITS statistic tend to increase with m and decrease with n.

DFBETAS statistics
: are standardized differences in the individual coefficient estimates resulting from the omission of the observation. They are identified in the column headings with the names of the independent variables. These statistics are primarily useful only for observations with large DFFITS values. Relatively large DFBETAS may indicate the independent variables that may be the cause of the problem. These statistics do not follow the *t* distribution; they tend to become smaller as sample size increases. Therefore, it is most useful to look for relatively large values within any specific row.

All of these statistics, with the exception of the DFBETAS, can be output to a data set. For this example, only the DFFITS statistics are requested. The keywords RSTUDENT, H, and COVRATIO are used to output the other statistics. One other statistic, called PRESS, which is not printed as part of the output, is available in the output data set using the keyword PRESS. The PRESS statistics give the residuals obtained by estimating the equation with all other observations (Allen 1970, 469-471).

Finally, the Cook's *D* statistic (available with the R option) is a version of the DFFITS statistic. Cook's *D* statistic measures the change in the estimates that results from deleting each observation. Essentially, Cook's *D* statistic is the DFFITS statistic, scaled and squared to make extreme values stand out more clearly. Furthermore, the all observations variance estimate is used for standardization.* You can also include this statistic in the output data set by using the keyword COOK.

The statistics printed as a result of the INFLUENCE option (**Output 3.6**) confirm the suspicions that all overall statistics and the DFBETAS for RATIO appear unusually large for the last observation. The plot of the DFFITS values against the predicted values (shown in **Output 3.7**) reinforces these impressions. The following SAS statements are required:

```
PROC PLOT;
   PLOT DFFITS*PPERC;
```

* The computation of the influence statistics does not require the recomputation of regression omitting each observation. All of these statistics are functions of the results of the original model and the h_i. Computational and other details can be found in Belsley et al. (1980).

Output 3.7 Plot of the DFFITS Statistics

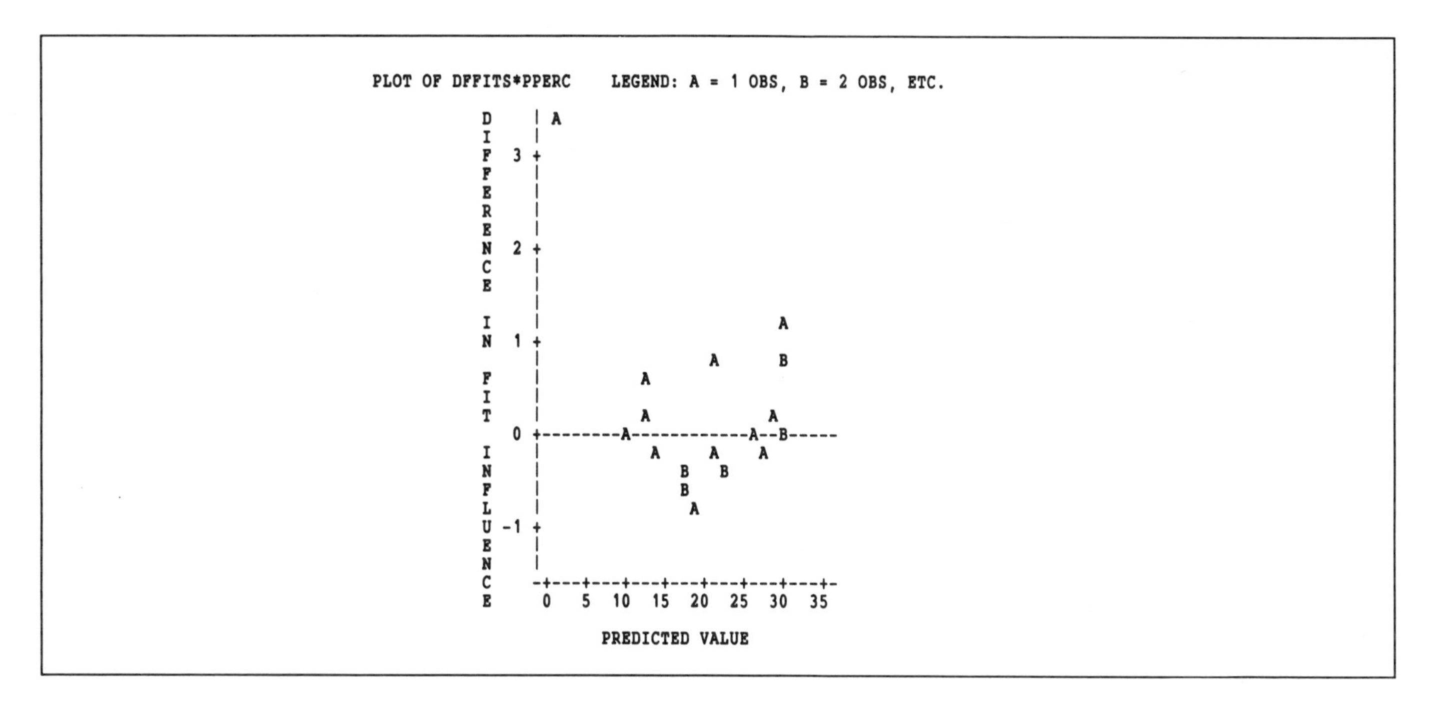

Further inspection of the original data quickly reveals the problem: the RATIO value for this observation appears out of range.

As previously noted, what you do now is not strictly a statistical problem. Discarding observations simply because they do not fit is bad statistical practice. Normally, the first thing to do is to check the original data to see if an error has been made. If the values are correct but represent an unusual situation that is not intended to be covered by the model, then the decision on the proper course of action must be made by an expert in the field of study, not by the statistician.

Although it is not necessarily the correct procedure, you can eliminate the last observation from the data set and resubmit the regression. The results are given in **Output 3.8**.

Output 3.8 Regression Results Omitting the Unusual Observation

```
DEP VARIABLE: PERC

                          ANALYSIS OF VARIANCE

                        SUM OF          MEAN
SOURCE       DF        SQUARES        SQUARE       F VALUE       PROB>F

MODEL         4     1305.30832     326.32708        26.175       0.0001
ERROR        16      199.47526   12.46720375
C TOTAL      20     1504.78358

     ROOT MSE       3.530893      R-SQUARE          0.8674
     DEP MEAN       21.44762      ADJ R-SQ          0.8343
     C.V.           16.46287
```

(continued on next page)

(continued from previous page)

PARAMETER ESTIMATES

VARIABLE	DF	PARAMETER ESTIMATE	STANDARD ERROR	T FOR H0: PARAMETER=0	PROB > |T|
INTERCEP	1	44.57305940	9.03179883	4.935	0.0001
RATIO	1	-6.47555196	0.76553517	-8.459	0.0001
INFT	1	-147.57022	81.73290588	-1.806	0.0898
LOST	1	1.92001643	1.09872683	1.747	0.0997
ADVT	1	0.38678986	7.05773023	0.055	0.9570

The changes in the estimated regression relationship are quite marked. The residual mean square has decreased considerably; the coefficient for RATIO has increased in magnitude, and none of the other variables are now significant at the 0.05 level.

Before continuing, it is important to point out that these statistics do not often provide clear-cut evidence of outliers. Not all of the different statistics are designed to detect the same type of data anomalies; therefore, these statistics may provide apparently contradictory results. Furthermore, the analysis may fail entirely. This situation can easily occur if there are several outliers. In other words, these influence statistics are only tools to aid in outlier detection; they cannot replace careful data monitoring.

3.3 SPECIFICATION ERRORS

You can also detect specification errors by examining the estimated residuals from a fitted equation. Patterns exhibited by the residuals usually reveal specification errors. Thus, there are few advantages for the use of those statistics designed primarily to identify single outliers. A curvilinear relationship is a common pattern suggested by the residuals, although other patterns such as bunching or cycling are possible. The residuals from the regression of the irrigation data, omitting the unusual observation, are plotted in **Output 3.9**. The following SAS statements are required:

```
PROC PLOT;
   PLOT RPERC*PPERC;
```

A curved pattern in the residuals is evident. This suggests that a curvilinear component, probably a quadratic term in one or more variables, should be included in the model. Unless there is some prior knowledge that would suggest which variable(s) should have the added quadratic term, you can use plots of residuals against each of the independent variables for that purpose.

Output 3.9 Residuals When One Observation Is Omitted

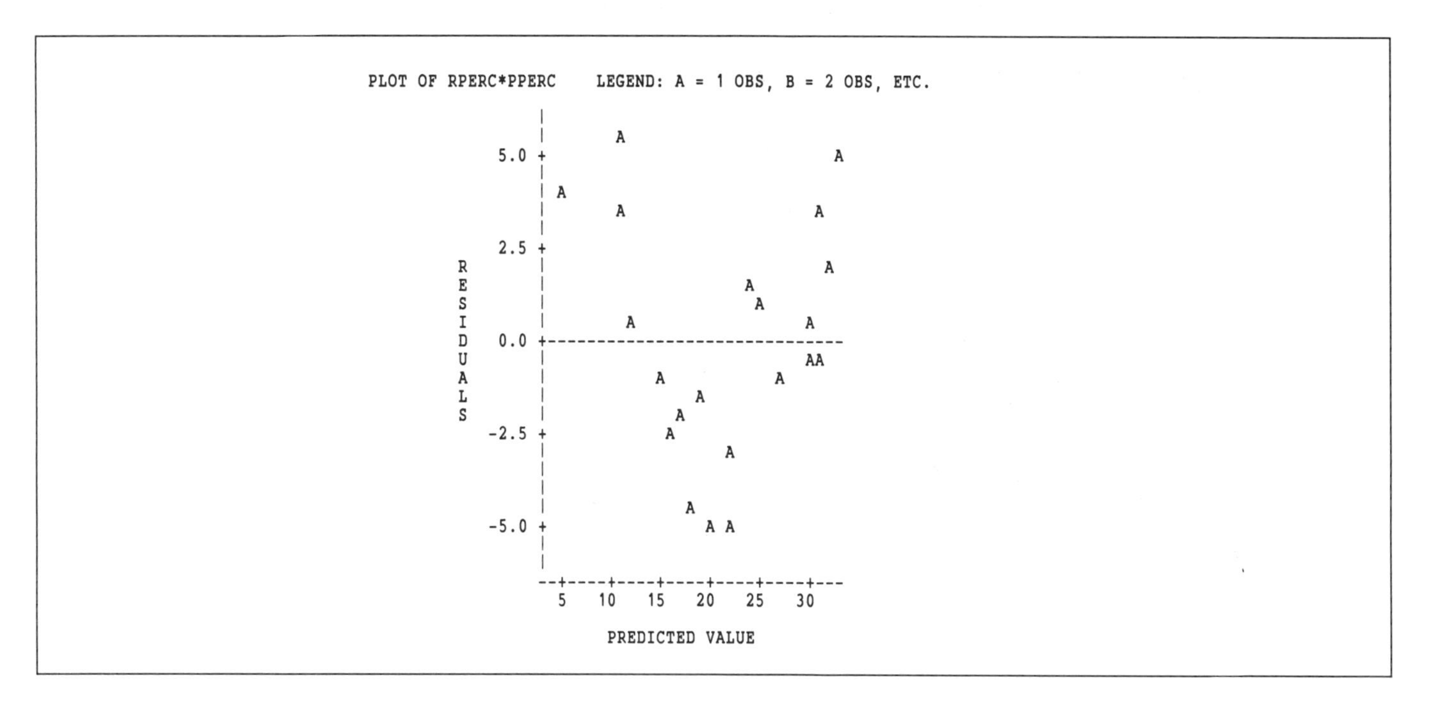

You can also accomplish this by examining the partial residual plots, also known as *leverage plots*. These are plots of the values of the dependent variable against the residuals of the regression of each independent variable on all other independent variables. The data points represented in these plots are the basis for the estimation of the partial regression coefficient; hence, these data points may reveal the need for curvilinear terms (see Freund and Minton 1979, 38). These plots are available with the PARTIAL option in the MODEL statement. Such plots are not very useful for outlier detection because the values used for the plots are not available. This makes identification of individual observations difficult. The ID statement causes the value of the ID variable to be used as a plotting symbol. This is primarily useful if the ID variable consists of single character values. These SAS statements create the plot:

```
PROC REG;
   MODEL PERC=RATIO INFT LOST ADVT / R INFLUENCE PARTIAL;
   OUTPUT OUT=B P=PPERC R=RPERC RSTUDENT=RSTUDENT;
   ID RATIO;
```

Output 3.10 gives the leverage plot for the variable RATIO. This plot shows a definite curved pattern which is not evident in the plots for the other variables. Hence, the addition of a quadratic term in the RATIO variable may be useful.

The following variable is created in the DATA step:

```
RATSQ=RATIO*RATIO;
```

Output 3.10 Partial Regression Residual Plot

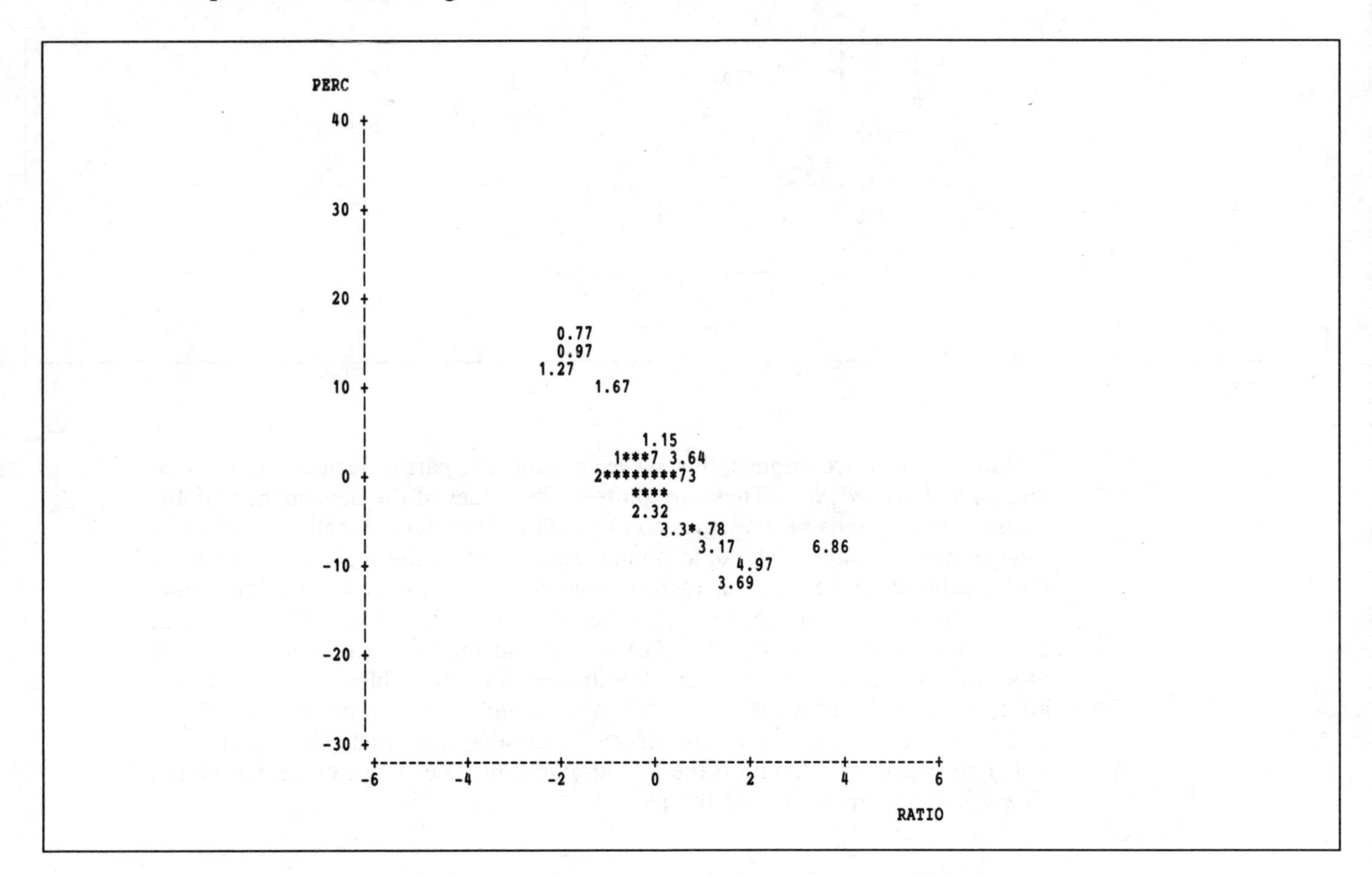

Now the regression is performed:

```
PROC REG;
   MODEL PERC=RATIO RATSQ INFT LOST ADVT;
```

These statements produce the results shown in **Output 3.11**.

Output 3.11 Regression Using RATIO and RATSQ

```
DEP VARIABLE: PERC
                                  ANALYSIS OF VARIANCE

                                SUM OF         MEAN
              SOURCE     DF    SQUARES       SQUARE       F VALUE       PROB>F

              MODEL       5   1409.48431    281.89686      44.370       0.0001
              ERROR      15  95.29926604   6.35328440
              C TOTAL    20   1504.78358

                 ROOT MSE     2.520572     R-SQUARE      0.9367
                 DEP MEAN     21.44762     ADJ R-SQ      0.9156
                 C.V.         11.75222

                                  PARAMETER ESTIMATES

                         PARAMETER        STANDARD       T FOR H0:
       VARIABLE  DF       ESTIMATE           ERROR     PARAMETER=0     PROB > |T|      TYPE I SS

       INTERCEP   1    56.38281754      7.07640680           7.968         0.0001     9660.00762
       RATIO      1   -15.78851897      2.36390617          -6.679         0.0001     1263.55655
       RATSQ      1     1.73829739      0.42927876           4.049         0.0010      121.04245
       INFT       1    -102.55434      59.39568226          -1.727         0.1048    10.79947942
       LOST       1     1.19532670      0.80449833           1.486         0.1580    14.08129368
       ADVT       1    -0.13481694      5.03989594          -0.027         0.9790    0.004546153
```

If you compare these results with those shown in **Output 3.8**, you can see that the residual mean square has been halved. In other words, the better fitting equation has been developed using data without the last observation. Thus, the relationship is one where PERC decreases with RATIO, but the rate of decrease diminishes with increasing values of RATIO. The other variables remain statistically insignificant; hence, you may consider variable selection (refer to Chapter 4).

However, it is of interest to note that the outlier detection statistics for this model (not reproduced here) indicate that the last observation may also be an outlier. This happens quite frequently; the elimination of one apparent outlier creates another. Such phenomena reinforce the danger of deleting observations simply because they do not fit.

3.4 HETEROGENEOUS VARIANCES

A fundamental assumption underlying linear regression analysis is that the random errors (the ε_i) have the same variance. Outliers can be considered a special case of unequal variances. Such observations may be considered to have very large variances.

Violations of the equal variance assumption are usually detected by residual plots that may reveal groupings of observations with large residuals suggesting larger variances. Some of the other outlier detection statistics can also be helpful, especially when the violation occurs in only a small number of observations.

In many applications there is a specific pattern for the magnitudes of variances. The most common pattern is an increase in variation for larger values of the response variable. For cases like this, using a transformation of the variable is in order (Steel and Torrie 1980). The most popular transformation, especially in regression analysis, is the logarithmic transformation that is discussed in Chapter 6. If the use of transformations is not appropriate, it is sometimes useful to alter the estimation procedure by either modifying the least-squares method or implementing a different estimation principle.

These methods are illustrated here using data on the sales prices of a set of investment grade diamonds. The data are used to estimate the relationship of the sales price (PRICE) to the weight (CARATS) of the diamonds. The following SAS statements produce the corresponding plot presented in **Output 3.13**.

```
PROC PLOT;
     PLOT PRICE*CARATS / HPOS=30 VPOS=25 VAXIS=0 TO 110000 BY 10000;
```

The data are given in **Output 3.12**.

Output 3.12 Diamond Data

CARATS	PRICE	CARATS	PRICE
0.50	1918	1.02	27264
0.52	2055	1.02	12684
0.52	1976	1.03	11372
0.53	1976	1.06	13181
0.54	2134	1.23	17958
0.60	2499	1.24	18095
0.63	2324	1.25	19757
0.63	2747	1.29	36161
0.68	2324	1.35	15297
0.73	3719	1.36	17432
0.75	5055	1.41	19176
0.77	3951	1.46	16596
0.79	4131	1.66	16321
0.79	4184	1.90	28473
0.91	4816	1.92	100411

Output 3.13 Plot of Diamond Data

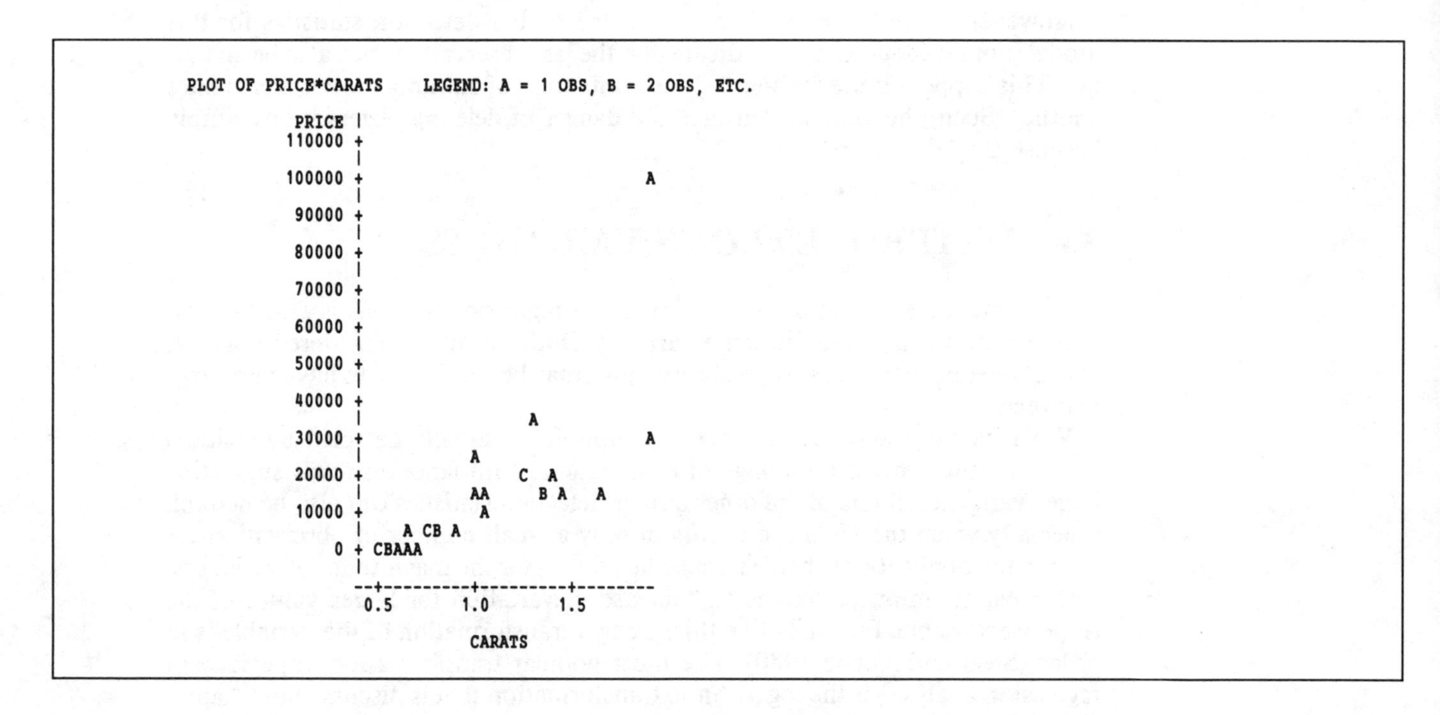

The plot clearly reveals the large variation of prices for larger diamonds. It is difficult to determine if the most expensive diamond is an outlier or simply a diamond that reflects the much higher variability of prices for larger stones. The plot also suggests an upward curving response; therefore, you implement a quadratic regression. In the DATA step the variable CSQ is created as the square of CARATS, and then the regression is implemented in the following PROC step:

```
PROC REG;
   MODEL PRICE=CARATS CSQ / R;
   OUTPUT OUT=A P=PQP;
```

The output is shown in **Output 3.14**.

Output 3.14 Quadratic Regression

```
DEP VARIABLE: PRICE
                                    ANALYSIS OF VARIANCE

                                  SUM OF          MEAN
              SOURCE     DF      SQUARES        SQUARE       F VALUE        PROB>F

              MODEL       2   6328150344    3164075172        21.590        0.0001
              ERROR      27   3956915101     146552411
              C TOTAL    29  10285065445

                 ROOT MSE       12105.88      R-SQUARE        0.6153
                 DEP MEAN       13866.23      ADJ R-SQ        0.5868
                 C.V.           87.30477

                                    PARAMETER ESTIMATES

                               PARAMETER        STANDARD       T FOR H0:
          VARIABLE   DF         ESTIMATE           ERROR     PARAMETER=0        PROB > |T|

          INTERCEP    1      11269.77056  15304.69264           0.736            0.4679
          CARATS      1     -30614.70351  29630.90647          -1.033            0.3107
          CSQ         1      28439.56106  12884.34421           2.207            0.0360


                       PREDICT     STD ERR                  STD ERR     STUDENT
  OBS     ACTUAL         VALUE     PREDICT    RESIDUAL     RESIDUAL    RESIDUAL      -2-1-0 1 2

    1     1918.0        3072.3      4486.4     -1154.3      11243.9     -0.1027      |     |     |
    2     2055.0        3040.2      4226.8      -985.2      11344.0     -0.0868      |     |     |
    3     1976.0        3040.2      4226.8     -1064.2      11344.0     -0.0938      |     |     |
    4     1976.0        3032.7      4103.5     -1056.7      11389.2     -0.0928      |     |     |
    5     2134.0        3030.8      3984.6      -896.8      11431.4     -0.0785      |     |     |
    6     2499.0        3139.2      3369.7      -640.2      11627.5     -0.0551      |     |     |
    7     2324.0        3270.2      3130.0      -946.2      11694.2     -0.0809      |     |     |
    8     2747.0        3270.2      3130.0      -523.2      11694.2     -0.0447      |     |     |
    9     2324.0        3602.2      2837.5     -1278.2      11768.7     -0.1086      |     |     |
   10     3719.0        4076.5      2676.7      -357.5      11806.3     -0.0303      |     |     |
   11     5055.0        4306.0      2646.1       749.0      11813.1      0.0634      |     |     |
   12     3951.0        4558.3      2632.6      -607.3      11816.2     -0.0514      |     |     |
   13     4131.0        4833.3      2634.0      -702.3      11815.8     -0.0594      |     |     |
   14     4184.0        4833.3      2634.0      -649.3      11815.8     -0.0550      |     |     |
   15     4816.0        6961.2      2839.5     -2145.2      11768.2     -0.1823      |     |     |
   16    27264.0        9631.3      3104.4     17632.7      11701.1      1.5069      |     |***  |
   17    12684.0        9631.3      3104.4      3052.7      11701.1      0.2609      |     |     |
   18    11372.0        9908.2      3125.6      1463.8      11695.4      0.1252      |     |     |
   19    13181.0       10772.9      3184.2      2408.1      11679.6      0.2062      |     |     |
   20    17958.0       16639.9      3358.7      1318.1      11630.6      0.1133      |     |     |
   21    18095.0       17036.2      3361.5      1058.8      11629.8      0.0910      |     |     |
   22    19757.0       17438.2      3363.8      2318.8      11629.2      0.1994      |     |     |
```

(continued on next page)

(continued from previous page)

```
23   36161.0   19103.1   3368.8    17057.9   11627.7    1.4670  |       |**     |
24   15297.0   21771.0   3374.1    -6474.0   11626.2   -0.5568  |      *|       |
25   17432.0   22235.6   3376.1    -4803.6   11625.6   -0.4132  |       |       |
26   19176.0   24643.7   3398.6    -5467.7   11619.0   -0.4706  |       |       |
27   16596.0   27194.1   3455.7   -10598.1   11602.2   -0.9135  |      *|       |
28   16321.0   38817.4   4393.5   -22496.4   11280.5   -1.9943  |    ***|       |
29   28473.0   55768.6   7534.8   -27295.6    9475.2   -2.8808  |  *****|       |
30    100411   57329.1   7890.9    43081.9    9180.7    4.6926  |       |******|
```

As expected, the regression is certainly significant. The estimated equation is

$$\text{PRICE} = 11270 - 30614(\text{CARATS}) + 28440(\text{CARATS})^2 \quad .$$

The coefficient for $(\text{CARATS})^2$ is statistically significant, confirming the upward curve of the relationship of price to carats. The coefficient for (CARATS) has no practical significance (see Chapter 5). The residuals, as well as the standardized residuals, clearly show the increasing price variation for the larger diamonds and suggest that these values may be unduly influencing the estimated regression relationship.

3.4.1 Weighted Least Squares

Weighted least squares is one principle that you can use to reduce influence of observations. Weighted least squares is a direct application of generalized least squares (Montgomery and Peck 1982, sec. 9.2). The estimated parameters obtained by this method are those that minimize the weighted residual sum of squares:

$$\Sigma w_i(y_i - \beta_0 - \beta_1 x_{1i} - \ldots - \beta_m x_{mi})^2$$

where w_i s are a set of nonnegative weights assigned to the individual observations. Observations with small weights contribute less to the sum of squares and, thus, provide less influence to the estimation of parameters, and vice versa for observations with larger weights. Thus, it is logical to assign small weights to observations whose large variances make them more unreliable and likewise to assign larger weights to observations with smaller variances. In fact, it can be shown that the best linear unbiased estimates are obtained if the weights are inversely proportional to the variances of the residuals.*

It is well known that prices tend to vary by proportions. This implies that standard deviations are proportional to means; hence, the variances are proportional to the squares of means. In sample data you do not know the values of the means; however, as a first approximation, you can use the observed prices as estimates of means and consequently use the reciprocal of the squares of prices as weights.

This principle is applied to the diamond prices data. Weighted regression is implemented in PROC REG using a WEIGHT statement that specifies the name of the variable to be used for the weights. This variable is created in the DATA step with the following statement:

```
W=1 / (PRICE*PRICE);
```

---

* This principle follows from the use of generalized least squares (Montgomery and Peck 1982, sec. 9.2).

The regression is implemented by PROC REG:

```
PROC REG;
   MODEL PRICE=CARATS CSQ / R;
   WEIGHT W;
   OUTPUT OUT=B P=PQWP;
```

The results are shown in **Output 3.15** (the Cook's D statistic is not shown).

Output 3.15 Weighted Regression

```
DEP VARIABLE: PRICE
                                   ANALYSIS OF VARIANCE

                                  SUM OF          MEAN
                SOURCE    DF     SQUARES        SQUARE       F VALUE       PROB>F

                MODEL      2   9.87697064   4.93848532        51.420       0.0001
                ERROR     27   2.59315502   0.09604278
                C TOTAL   29  12.47012566

                   ROOT MSE    0.3099077     R-SQUARE         0.7921
                   DEP MEAN      2768.42     ADJ R-SQ         0.7766
                   C.V.       0.01119439
                                   PARAMETER ESTIMATES

                           PARAMETER        STANDARD       T FOR H0:
          VARIABLE  DF      ESTIMATE           ERROR     PARAMETER=0      PROB > |T|

          INTERCEP   1    2071.63682      2459.54488           0.842          0.4070
          CARATS     1   -6960.71024      6286.96386          -1.107          0.2780
          CSQ        1   12647.35750      3689.12666           3.428          0.0020


                         PREDICT     STD ERR                 STD ERR    STUDENT
  OBS        ACTUAL        VALUE     PREDICT   RESIDUAL     RESIDUAL   RESIDUAL   -2 -1  0  1  2

    1        1918.0       1753.1       299.0      164.9        513.7     0.3209   |     |     |
    2        2055.0       1871.9       263.0      183.1        580.0     0.3157   |     |     |
    3        1976.0       1871.9       263.0      104.1        553.0     0.1882   |     |     |
    4        1976.0       1935.1       248.1    40.8969        559.9     0.0730   |     |     |
    5        2134.0       2000.8       235.4      133.2        618.0     0.2155   |     |     |
    6        2499.0       2448.3       213.2    50.7406        744.5     0.0682   |     |     |
    7        2324.0       2706.1       232.2     -382.1        681.8    -0.5605   |    *|     |
    8        2747.0       2706.1       232.2    40.8744        819.0     0.0499   |     |     |
    9        2324.0       3186.5       285.3     -862.5        661.3    -1.3042   |   **|     |
   10        3719.0       3730.1       345.7   -11.0952       1099.5    -0.0101   |     |     |
   11        5055.0       3965.2       369.6     1089.8       1522.4     0.7158   |     |*    |
   12        3951.0       4210.5       392.8     -259.5       1159.7    -0.2238   |     |     |
   13        4131.0       4465.9       415.3     -334.9       1211.0    -0.2765   |     |     |
   14        4184.0       4465.9       415.3     -281.9       1228.3    -0.2295   |     |     |
   15        4816.0       6210.7       536.8    -1394.7       1392.7    -1.0014   |   **|     |
   16       27264.0       8130.0       647.3    19134.0       8424.5     2.2712   |     |**** |
   17       12684.0       8130.0       647.3     4554.0       3877.2     1.1745   |     |**   |
   18       11372.0       8319.7       658.6     3052.3       3462.2     0.8816   |     |*    |
   19       13181.0       8903.9       694.8     4277.1       4025.4     1.0625   |     |**   |
   20       17958.0      12644.2       995.3     5313.8       5475.6     0.9705   |     |*    |
   21       18095.0      12886.9      1019.4     5208.1       5514.4     0.9445   |     |*    |
   22       19757.0      13132.2      1044.2     6624.8       6033.1     1.0981   |     |**   |
   23       36161.0      14138.8      1151.8    22022.2      11147.2     1.9756   |     |***  |
   24       15297.0      15724.5      1338.1     -427.5       4547.9    -0.0940   |     |     |
   25       17432.0      15997.6      1372.1     1434.4       5225.2     0.2745   |     |     |
   26       19176.0      17401.2      1554.6     1774.8       5735.9     0.3094   |     |     |
   27       16596.0      18868.1      1758.0    -2272.1       4833.5    -0.4701   |     |     |
   28       16321.0      25367.9      2776.5    -9046.9       4227.8    -2.1398   | ****|     |
   29       28473.0      34503.2      4416.6    -6030.2       7639.2    -0.7894   |    *|     |
   30        100411      35330.3      4573.3    65080.7      30780.2     2.1144   |     |**** |
```

A comparison of the results with those of the unweighted least-squares regression (**Output 3.14**) shows the effect of weighting. Since all sums of squares reflect the weights, they cannot be compared with the sums of squares of the unweighted analysis. However, the R-SQUARE statistic may be comparable since it is a ratio, and the effect of the weights is canceled. The R-SQUARE for the weighted analysis is somewhat larger because the effect of the largest residuals has been reduced. The estimated equation has a smaller coefficient for $(\text{CARATS})^2$. In other words, the curve has a somewhat smaller upward curvature that is presumably due to the lesser influence of the very high-priced diamonds. The two equations can be compared by using the following statements to create a plot:

```
DATA ALL;
   MERGE A B;
PROC PLOT;
   PLOT PRICE*CARATS='X' PQP*CARATS='U' PQWP*CARATS='W' / OVERLAY;
```

As in several previous examples, you can use SAS/GRAPH software's GPLOT procedure to provide a clearer plot. Use the following statements:

```
PROC GPLOT;
   PLOT PQP*CARATS=1
        PQWP*CARATS=2
        PRICE*CARATS=3 / OVERLAY VAXIS=0 TO 110000 BY 10000;
   SYMBOL1 V=U I=SPLINE C=BLACK;
   SYMBOL2 V=W I=SPLINE C=BLACK;
   SYMBOL3 V=SQUARE I=NONE C=BLACK;
```

The plot produced with these statements is shown in **Output 3.16**. This plot clearly shows a difference between the two estimated curves, but the difference is not great. It appears that the weighting was not very effective in reducing the influence of the most expensive diamonds.

Other weighting alternatives exist. For example, the estimated values of the dependent variable from a preliminary analysis can be used as the basis for the weights. For this example, the predicted values from the unweighted analysis (PQP) can be used to create the weight variable W=1/(PQP*PQP). This is then used as the weight variable in the regression analysis. The estimates of this analysis can be used in a second iteration as weights in a third estimation, and so forth. However, in the present example such an iterative procedure would increase the influence of the most expensive stone because its estimated price is lower (hence creating a relatively larger weight) than the actual price.

Another alternative weighting procedure that does not make any assumption concerning the pattern of variances is the use of residuals from a preliminary regression as the basis for the weights. This procedure assumes that the squares of the individual residuals reflect the variances of the corresponding residuals. Obviously, this procedure can also be iterated.

Output 3.16 Plot Produced with SAS/GRAPH Software

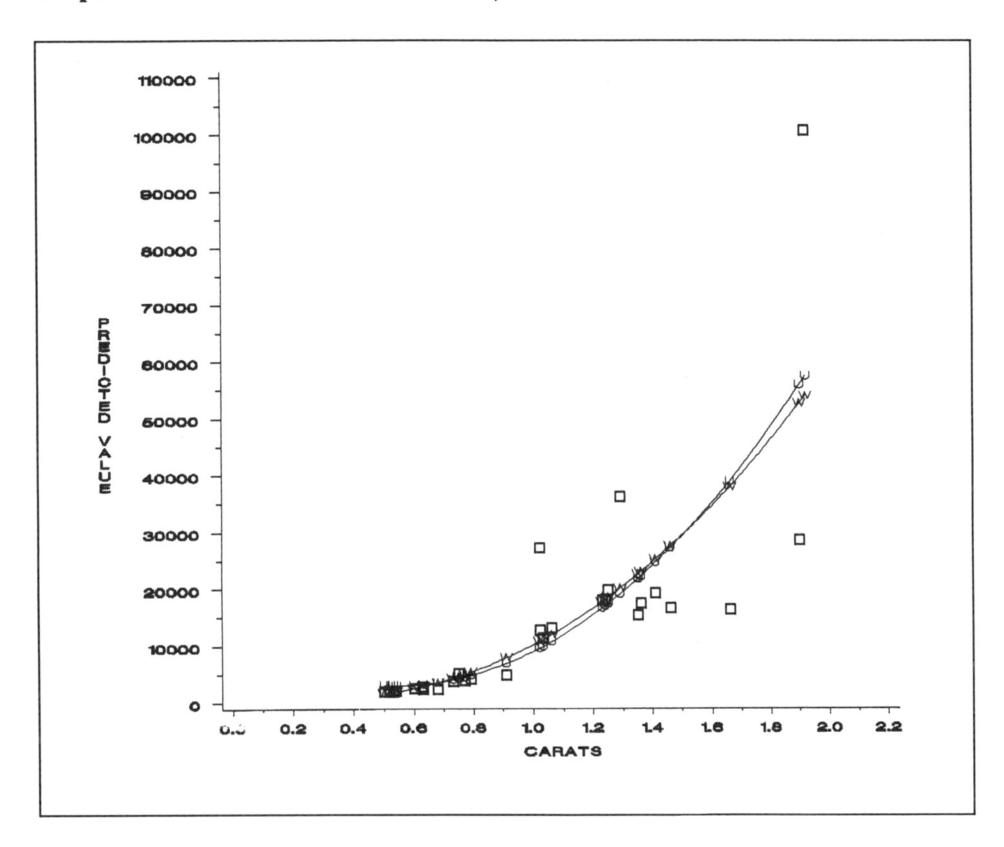

3.4.2 Least-Absolute-Deviations Estimation

Alternatives to weighted least squares include procedures using completely different estimation principles. These procedures are often referred to as *robust procedures.* One of these procedures minimizes the sum of absolute deviations by finding those estimates that minimize the value of

$$\Sigma \; | y_i - \beta_0 - \beta_1 x_{i1} - \ldots - \beta_m x_{im} | \quad .$$

This least-absolute-values criterion (also known as LAV, L1, MAD, MSAE, and other terms) is more difficult to compute and also suffers from the fact that little is known about sampling distributions of the resulting estimates. Thus, this criterion precludes the usual inference statistics. PROC LAV implements the LAV principle in the SAS System. PROC LAV is documented in the *SUGI Supplemental Library User's Guide, Version 5 Edition.*

You can use PROC LAV to analyze the diamond price data by using the following statements:

```
PROC LAV DATA=DIAMONDS OUT=F;
   MODEL PRICE=CARATS CSQ;
   OUTPUT PREDICTED=PQLP;
```

Note that in the PROC LAV procedure the name of the output data set (if required) is specified in the PROC statement rather than in the OUTPUT statement. Also, abbreviating the keywords in the OUTPUT statement is not permitted. The results are shown in **Output 3.17**.

Output 3.17 Using PROC LAV

```
        VARIABLE        LAV COEFFICIENT

            INTER       10347.12349639
            CARATS      21861.84977279
            CSQ          -752.73990912
     (NOTE: THE COEFFICIENT ESTIMATES ARE UNIQUE.)

       RESIDUAL SUM OF ABSOLUTE VALUES =        154741.32798717
       ADJUSTED TOTAL SUM OF ABSOLUTE VALUES = 324369.00000000
       NUMBER OF OBSERVATIONS IN DATA SET =         30
```

The output from PROC LAV is much shorter than that from PROC REG since there are no formal inference statistics. The estimated coefficients, labeled LAV COEFFICIENT, show that this estimation method produces a very different curve: the quadratic coefficient is negative, albeit rather small. In other words, the influence of the large diamonds has been reduced, and the relationship of price to carats no longer shows an increasing rate. The only other statistics give the total sum of absolute deviations from the mean and the sum of absolute values of the residuals. These statistics provide some measure of the goodness of fit, but they cannot be compared to the usual total and residual sums of squares.

Again use PROC GPLOT to compare the curves predicted by least squares and the LAV procedure:

```
PROC GPLOT;
   PLOT PQP*CARATS=1
        PQLP*CARATS=2
        PRICE*CARATS=3 / OVERLAY VAXIS=0 TO 110000 BY 10000;
   SYMBOL1 V=U I=SPLINE C=BLACK;
   SYMBOL2 V=W I=SPLINE C=BLACK;
   SYMBOL3 V=SQUARE I=NONE C=BLACK;
```

The plot is shown in **Output 3.18.**

The difference is quite evident; the negative quadratic component is of little importance and can probably be deleted from the model with little loss of precision. However, there are no statistics to support this conjecture; the regression must be recomputed with that term omitted to obtain the effect of dropping that term. However, it is questionable whether the LAV method has provided a better fit to the data since the estimated prices for many of the lower-priced diamonds are unrealistically low.

Output 3.18 A Comparison of Curves Using PROC GPLOT

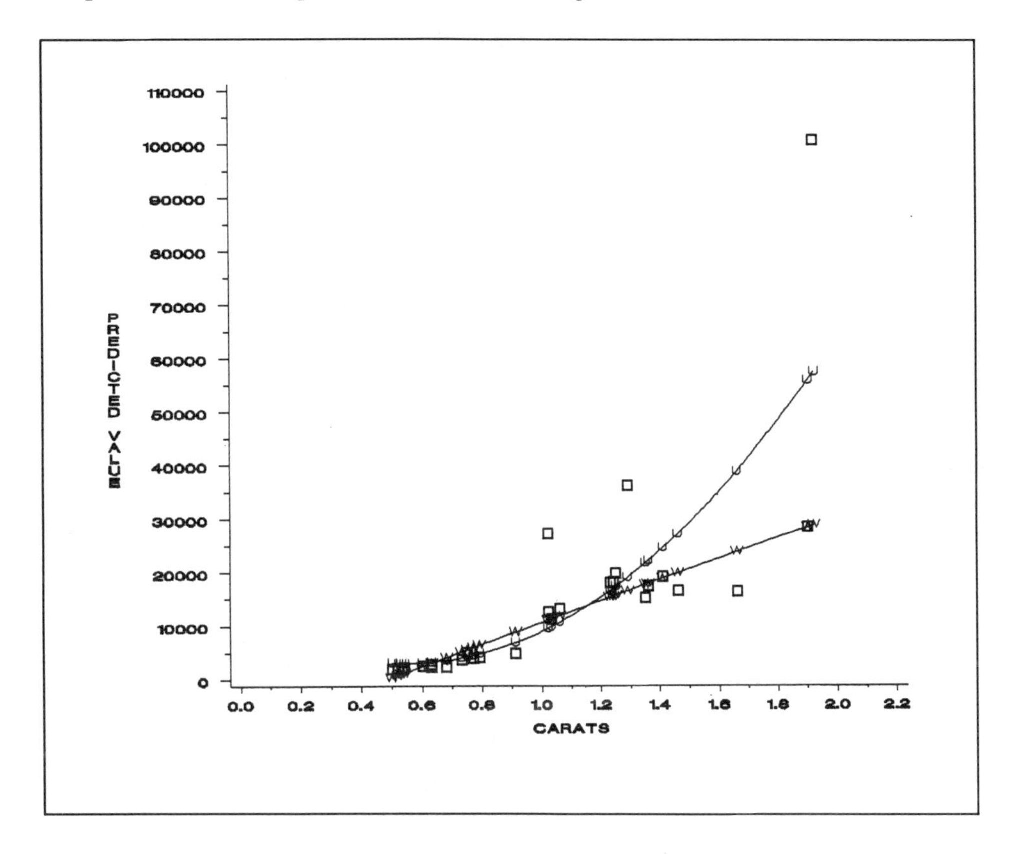

Of course, other robust estimation principles do exist. Minimization of other powers of residuals can be implemented by using iteratively reweighted least squares using PROC NLIN. (For example, see **Example 5** in "The NLIN Procedure" in the *SAS User's Guide: Statistics, Version 5 Edition.*)

You can also analyze this set of data using a log-linear model like the one described in Chapter 6, where the logarithmic transformation reduces the influence of the high-priced diamonds. Finally, it must be emphasized that there is no universal criterion for the best procedure for overcoming failures of assumptions. As always, you must know your data and choose the appropriate statistical method.

3.5 CORRELATED ERRORS

Another assumption underlying the distribution of the random errors is that the errors be independent. That is, the value of any one error is independent of the value of any other. Most violations of this assumption occur in time series data, that is, data observed over a sequence of time periods. In such data the observed value at a given point in time, say in period t, may be influenced by values observed in previous periods. For example, the weather today is highly influenced by what the weather was yesterday. Nonindependent errors can occur in other situations; for example, plants that are too close to each other may compete to the extent that larger plants may be surrounded by smaller, stunted plants.

In linear models, nonindependent errors are usually described as being correlated. If these correlations are known, you can use generalized least squares to provide correct estimates and other statistics. Not only are these correlations not known, but they cannot even be estimated in the general case since there are (N)(N−1)/2 correlations. This number greatly exceeds the number of observations (N) on which such estimates must be based.

Imposing a structure on the correlations obviates this problem somewhat. The most commonly used structure is the *autoregressive process*. This process specifies that the error at time *t* is related to previous errors by a linear regression model. The number of previous errors in this model is called the order of the autoregressive process. The most commonly used order is one; that is, the errors at time *t* are directly related only to the errors of the previous period.

It can be shown that if the existence of an autoregressive process is ignored, that is, if ordinary least squares is used for estimation and inference, then the estimates of coefficients are unbiased, but estimates of the residual variance and of the regression coefficients are subject to an unknown bias. Therefore, it is necessary to consider alternative analysis methodologies.*

This section presents an example to illustrate one procedure for detecting the existence of a first order autoregressive model. Also, this section presents an alternative analysis methodology provided by PROC AUTOREG, a SAS procedure documented in the *SAS/ETS User's Guide, Version 5 Edition* and in *SAS System for Forecasting Time Series, 1986 Edition.*

A consumption function is a model that attempts to estimate consumption of goods and services using other economic variables. From the quarterly data on consumption (CONS), shown in **Output 3.19**, you want to estimate a model using the following variables:

CURR	is currency in circulation.
DDEP	is the amount of demand deposits.
GNP	is the gross national product.
WAGES	is the hourly wage rate.
INCOME	is the national income.

All variables have been deflated by the consumer price index (January, 1953=100), and all variables except WAGES have been converted to a per-capita basis. The data consist of seventy-six observations from the first quarter of 1951 through the last quarter of 1969. The data are given in **Output 3.19**.

Output 3.19 Quarterly Data on Consumption

YR	QTR	CURR	DDEP	GNP	WAGES	CONS
51	1	181.3	664.5	2296.9	1.713	1256.1
51	2	181.1	665.0	2332.2	1.739	1247.7
51	3	181.7	664.0	2353.4	1.736	1254.5
51	4	180.7	666.3	2340.8	1.746	1256.9
52	1	183.2	677.2	2365.2	1.779	1265.9
52	2	182.3	674.3	2333.1	1.764	1276.3
52	3	182.3	673.3	2350.5	1.796	1287.5
52	4	184.2	676.4	2422.1	1.828	1306.8
53	1	186.3	678.2	2467.4	1.868	1321.1
53	2	185.6	674.6	2461.8	1.865	1317.7
53	3	184.2	668.0	2424.2	1.874	1309.5
53	4	183.9	668.1	2386.8	1.891	1307.9
54	1	182.6	669.0	2377.3	1.891	1320.1

(continued on next page)

* The analysis of time series data is a major specialty within the discipline of statistics. SAS/ETS software contains some of the most popular methods for analyzing time series data.

(continued from previous page)

54	2	180.7	666.6	2360.1	1.898	1322.8
54	3	179.8	674.7	2384.2	1.904	1338.2
54	4	178.9	683.1	2438.5	1.942	1355.1
55	1	178.8	689.3	2511.4	1.953	1367.5
55	2	178.7	690.9	2553.9	1.974	1379.2
55	3	177.7	687.9	2582.7	2.009	1382.2
55	4	177.7	685.9	2613.4	2.032	1408.3
56	1	177.6	685.6	2614.0	2.053	1422.2
56	2	174.6	676.0	2605.2	2.059	1414.0
56	3	173.2	667.9	2601.1	2.075	1416.8
56	4	171.5	661.7	2621.6	2.100	1419.8
57	1	170.2	655.4	2636.7	2.095	1428.5
57	2	167.6	646.0	2614.3	2.082	1416.8
57	3	166.2	637.9	2621.3	2.087	1429.0
57	4	164.8	628.3	2570.7	2.099	1427.7
58	1	161.3	616.5	2485.8	2.070	1410.8
58	2	160.8	621.0	2490.4	2.083	1423.9
58	3	160.6	625.6	2553.1	2.103	1439.5
58	4	160.5	631.7	2614.7	2.153	1447.0
59	1	161.0	635.4	2658.2	2.173	1464.8
59	2	160.3	633.0	2701.0	2.177	1469.0
59	3	159.4	629.8	2659.8	2.147	1478.3
59	4	157.6	619.7	2675.6	2.190	1482.1
60	1	157.5	609.9	2732.0	2.207	1494.2
60	2	155.8	598.6	2712.0	2.192	1505.1
60	3	154.8	597.5	2692.2	2.197	1498.3
60	4	153.3	592.0	2660.1	2.204	1500.5
61	1	152.2	593.4	2651.5	2.204	1508.5
61	2	151.4	595.8	2698.1	2.231	1513.8
61	3	151.0	592.9	2719.2	2.218	1514.2
61	4	152.0	597.1	2779.8	2.268	1532.9
62	1	152.2	594.6	2807.9	2.267	1536.7
62	2	152.8	592.0	2838.8	2.270	1544.7
62	3	152.0	583.0	2841.6	2.253	1546.7
62	4	152.9	585.5	2877.3	2.297	1569.4
63	1	154.4	587.5	2884.3	2.298	1577.5
63	2	155.2	587.5	2896.6	2.308	1580.7
63	3	156.3	587.3	2922.5	2.306	1591.2
63	4	157.4	587.7	2952.3	2.333	1588.3
64	1	159.2	587.8	2998.1	2.331	1619.2
64	2	160.7	587.7	3030.1	2.343	1631.3
64	3	161.9	592.0	3060.2	2.362	1659.2
64	4	162.1	593.9	3067.4	2.371	1663.7
65	1	163.8	593.6	3137.2	2.376	1677.7
65	2	163.1	590.1	3155.6	2.371	1698.3
65	3	165.1	594.0	3209.3	2.387	1720.0
65	4	166.2	597.5	3261.1	2.396	1742.4
66	1	167.0	600.2	3303.4	2.393	1758.9
66	2	168.0	599.8	3317.4	2.400	1771.0
66	3	167.9	587.7	3326.2	2.401	1775.9
66	4	168.3	581.5	3357.4	2.415	1773.6
67	1	169.6	582.9	3358.6	2.426	1800.9
67	2	169.5	586.0	3359.4	2.431	1803.9
67	3	169.1	595.4	3386.8	2.434	1808.1
67	4	169.5	596.1	3450.5	2.462	1807.0
68	1	170.3	595.2	3486.3	2.477	1841.4
68	2	171.2	597.1	3533.6	2.481	1845.6
68	3	172.1	602.0	3557.5	2.488	1863.6
68	4	172.4	601.2	3570.0	2.514	1857.6
69	1	173.4	600.4	3572.8	2.492	1862.1
69	2	173.0	595.3	3570.3	2.484	1861.6

The regression is implemented as follows:

```
PROC REG;
   MODEL CONS=CURR DDEP GNP WAGES INCOME / DW;
   OUTPUT OUT=A P=PC R=RC;
```

The DW option in the MODEL statement requests the computation of the Durbin-Watson *d* statistic for establishing the existence of a first order autoregressive process. The output of a data set containing the predicted and residual values is also requested. The output of the procedure is shown in **Output 3.20**.

Output 3.20 Regression Analysis with the DW Option

```
DEP VARIABLE: CONS
                              ANALYSIS OF VARIANCE

                            SUM OF            MEAN
         SOURCE     DF     SQUARES          SQUARE       F VALUE        PROB>F

         MODEL       5  2702413.97       540482.79      3366.924        0.0001
         ERROR      70 11236.90328       160.52719
         C TOTAL    75  2713650.87

             ROOT MSE     12.66993       R-SQUARE         0.9959
             DEP MEAN     1532.125       ADJ R-SQ         0.9956
             C.V.        0.8269516

                              PARAMETER ESTIMATES

                        PARAMETER        STANDARD       T FOR H0:
   VARIABLE  DF          ESTIMATE           ERROR     PARAMETER=0      PROB > |T|

   INTERCEP   1         160.38765     95.79100505           1.674          0.0985
   CURR       1        1.48288870      0.60086544           2.468          0.0160
   DDEP       1       -0.57537163      0.14214897          -4.048          0.0001
   GNP        1        0.11587332      0.11466272           1.011          0.3157
   WAGES      1         323.56479     55.09252701           5.873          0.0001
   INCOME     1        0.19255673      0.11830923           1.628          0.1081

   VARIABLE  DF

   INTERCEP   1
   CURR       1
   DDEP       1
   GNP        1
   WAGES      1
   INCOME     1

DURBIN-WATSON D                0.949
(FOR NUMBER OF OBS.)              76
1ST ORDER AUTOCORRELATION      0.518
```

The model obviously fits the data quite well. The most important coefficients are those for demand deposits and wages. The significant negative coefficient for demand deposits is puzzling. There is some evidence of multicollinearity, and possibly variable selection is in order. This will be covered in Chapter 4.

The last two items in the output concern autocorrelation. The Durbin-Watson *d* statistic is a test for existence of a first order autoregressive process. According to the table for this statistic (see Montgomery and Peck 1982, Table A6; algorithms for computing *p* values are not available), a value of less than 1.25 (for n=60, m=5) indicates the existence of a positive first order autocorrelation at a significance level of less than 0.01. The second item is the actual sample correlation of adjacent residuals; the value of 0.518 shows a rather large, but not extreme, degree of association between adjacent residuals. This first order autocorrelation is also known as the *lag one* autocorrelation because it is the correlation of observations with those observations that lag behind one period.

A plot of residuals against time may also be useful in detecting autocorrelation. To do this, first create a time variable in the DATA step:

```
N=_N_;
```

This variable uses the automatic observation counter, _N_, to create the sequential period indicator. The plot is implemented as follows:

```
PROC PLOT;
   PLOT RC*N / VREF=0;
```

A plot of this type is more useful if each column of the plot represents one period; an HAXIS option may be needed to assure that the plot is constructed in this manner. The plot is shown in **Output 3.21**.

Output 3.21 Plot of Predicted and Residual Value

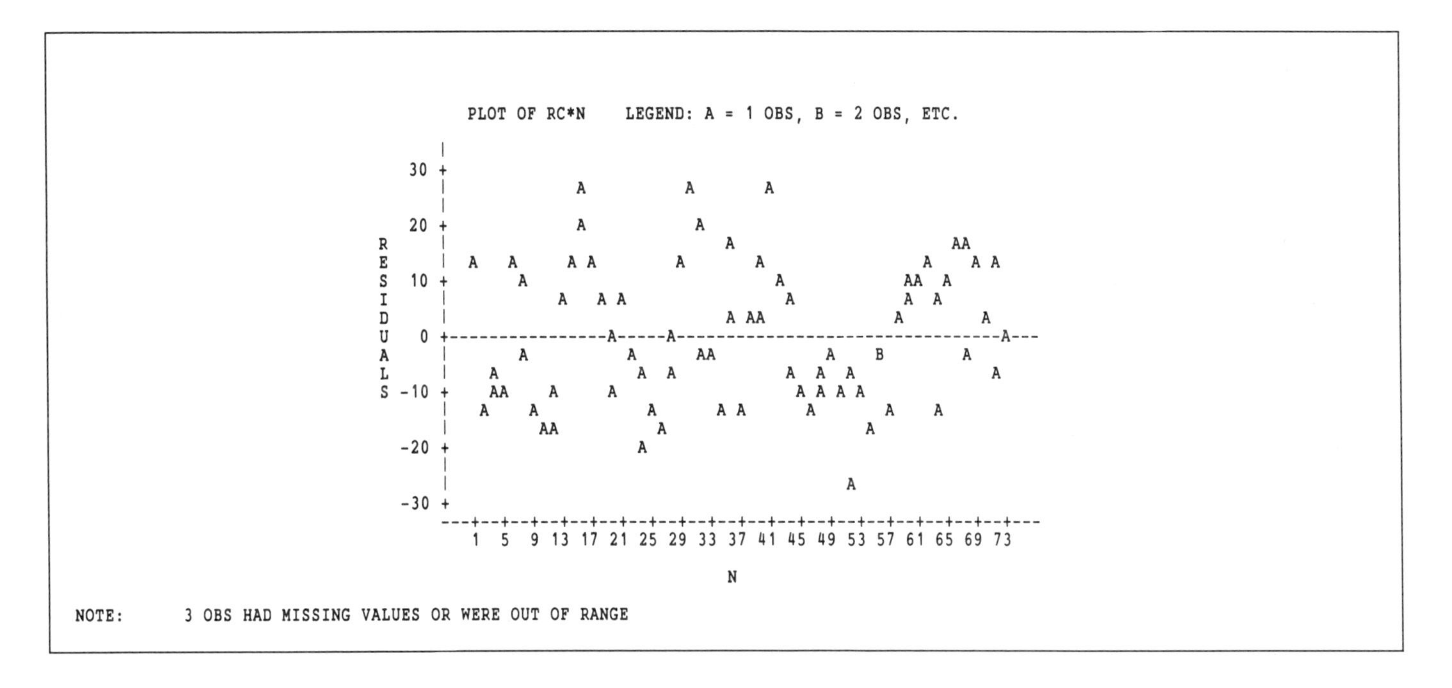

The plot appears to show cycles, that is, a number of sequences of either positive or negative residuals. This is due to the fact that under a first order autoregressive process with a positive autocorrelation, a positive residual is more likely to be followed by another positive residual and vice versa. Of course, true cyclical effects, such as seasons, can also cause apparent cycles in the residuals.

One popular alternative analysis procedure for data of this type is PROC AUTOREG. This procedure uses the residuals from an ordinary least-squares analysis to estimate the set of autoregressive parameters. You specify the dependent variable and the independent regressor variables in the MODEL statement. These coefficients are used to perform the appropriate generalized least-squares method that is implemented by performing a transformation of the variables in the model (see Fuller 1978, sec. 2.5). The procedure is implemented as follows:

```
PROC AUTOREG;
   MODEL CONS=CURR DDEP GNP WAGES INCOME / NLAG=4;
```

The statements are in the same form as PROC REG. The only other specification needed at this point is the NLAG= option in the MODEL statement. By specifying NLAG=4, you indicate that the autoregressive process involves lags up to and including four. This order is chosen to accommodate the possibility of a yearly (four quarters) cycle. Other MODEL and PROC options are discussed below. The output of this procedure is shown in **Output 3.22**.

Output 3.22 Using PROC AUTOREG

```
DEPENDENT VARIABLE = CONS


                              ORDINARY LEAST SQUARES ESTIMATES

                VARIABLE DF        B VALUE        STD ERROR   T RATIO APPROX PROB

                INTERCPT  1     160.387654       95.7910050     1.674      0.0985
                CURR      1       1.482889        0.6008654     2.468      0.0160
                DDEP      1      -0.575372        0.1421490    -4.048      0.0001
                GNP       1       0.115873        0.1146627     1.011      0.3157
                WAGES     1     323.564790       55.0925270     5.873      0.0001
                INCOME    1       0.192557        0.1183092     1.628      0.1081

                              ESTIMATES OF AUTOCORRELATIONS

            LAG  COVARIANCE  CORRELATION  -1 9 8 7 6 5 4 3 2 1 0 1 2 3 4 5 6 7 8 9 1
             0      147.854     1.000000   |                    |********************|
             1      76.5514     0.517750   |                    |**********          |
             2       39.698     0.268494   |                    |*****               |
             3     -2.29513    -0.015523   |                    |                    |
             4     -19.0695    -0.128975   |                 ***|                    |

                              PRELIMINARY MSE=        103.026

                        ESTIMATES OF THE AUTOREGRESSIVE PARAMETERS
                        LAG     COEFFICIENT      STD ERROR          T RATIO
                          1     -0.50517162     0.12287967        -4.111108
                          2     -0.11644384     0.13590145        -0.856826
                          3      0.18108645     0.13590145         1.332484
                          4      0.05864028     0.12287967         0.477217


                                  YULE-WALKER ESTIMATES

                        SSE            7618.971     DFE                  66
                        MSE            115.4389     ROOT MSE       10.74425
                        SBC            609.6314     AIC            586.3241
                        REG RSQ          0.9926     TOTAL RSQ        0.9972

                              A U T O R E G   P R O C E D U R E

                VARIABLE DF        B VALUE        STD ERROR   T RATIO APPROX PROB

                INTERCPT  1     262.076353       123.929923     2.115      0.0382
                CURR      1       0.922011         0.822528     1.121      0.2664
                DDEP      1      -0.517861         0.204377    -2.534      0.0137
                GNP       1       0.156508         0.143347     1.092      0.2789
                WAGES     1     261.044133        70.781143     3.688      0.0005
                INCOME    1       0.183093         0.147601     1.240      0.2192
```

The first portion of the output reproduces the coefficients estimated by ordinary least squares; the coefficients are the same as given in **Output 3.20**. No other statistics for this regression analysis are given. The residuals from this regression are used to compute the autocorrelations for the lags (specified by the NLAG= option) that are printed under the heading ESTIMATES OF AUTOCORRELATIONS. The coefficient for LAG 1 is the same as shown in the output for PROC REG. The plot of the autocorrelations provides a quick check on possible patterns of the various autocorrelations. For example, if there is only a first order autocorrelation with a correlation of ρ, then it can be shown that the second (LAG 2) autocorrelation is ρ^2, the third order correlation is ρ^3, and so forth. For a positive first order process, the plot would show the correlations tending rapidly to zero with increasing lag as they do here.

The autoregressive coefficients appear next in the output together with their standard deviations (standard errors) and *t* ratios, the ratio of the estimate to the standard error. These statistics confirm that there appears to be only a first order autoregressive process. Thus, these statistics suggest that the model should be re-estimated by specifying 1 in the NLAG= option.

These autoregressive parameter estimates are used to perform the transformations of the variables required to obtain the generalized least-squares estimates. The transformed variables are not printed but are available in an output data set. The resulting estimates make up the final portion of the output. A comparison of these estimates with those of the ordinary least-squares method shows that these estimates are not very different. Since the upward bias in variance estimates for a first order autoregressive model is generally accepted to be in the neighborhood of $1/(1-\rho^2)$, these results are not surprising. In other words, if the first order autocorrelation is 0.5, variance estimates using ordinary least squares should be approximately 25% too low.

PROC AUTOREG has a number of PROC and MODEL options relating to optional output of various intermediate results. MODEL options are used to restrict the number of autoregressive parameters to be estimated and also to perform, if desired, a backward elimination of autoregressive parameters. You can also request this procedure to produce an output data set containing predicted and residual values, the parameter estimates, and the transformed variables. These transformed variables can be used as input for repeated implementations of PROC AUTOREG, thus providing an iterative procedure for finding better estimates.

4. Multicollinearity: Detection and Remedial Measures

4.1 INTRODUCTION

In statistical analysis, you often create a model containing a large number of independent variables and assume that the data analysis will reveal the model you should have chosen initially. This approach suffers from two major problems:

1. The model and associated hypotheses are generated by the data, thus invalidating significance levels (p values). This problem is related to the control of Type I errors in multiple comparisons, and although the problem has been partially solved for multiple comparisons, the problem has not been solved for regression. Therefore, p values provided by many of the computer outputs in this chapter cannot be taken literally.
2. The inclusion of a large number of variables in a regression model often results in multicollinearity, which is defined as a high degree of multiple correlation among several independent variables.* This multicollinearity occurs because too many variables have been put into the model, and a number of these variables may measure similar phenomena.

The existence of multicollinearity is not a violation of the assumptions underlying the use of regression analysis. However, depending on the purpose of the analysis, multicollinearity may inhibit the usefulness of the results in the following ways:

1. The existence of multicollinearity does not affect the estimation of the dependent variable. In other words, the $\hat{y}$ values are the best linear unbiased estimates of the conditional means of the population.
2. The existence of multicollinearity tends to inflate the variances of predicted values, that is, the predictions of the response variable for sets of x values that are not included in the sample.
3. The existence of multicollinearity tends to inflate the variances of the parameter estimates. Multicollinearity often results in coefficient

---

* Correlations among independent variables also occur in polynomial models (Chapter 5) where, for example, there may be a positive correlation of x and x^2. This type of multicollinearity is not covered in this chapter.

estimates that are not statistically significant or have incorrect signs or magnitude (see Myers 1986, sec. 3.7). This is an especially troublesome result when it is important to ascertain the structure of the relationship of the response to the various independent variables.

Since conditions in (2) and (3) are essential for useful regression analyses, you must ascertain if multicollinearity exists. Then, if you determine that multicollinearity exists, it is important to determine the nature of the multicollinearity, that is, the nature of the linear relationships among the independent variables. The final step is to combat the effects of multicollinearity.

Section **4.2** discusses several methods available in PROC REG for the detection of multicollinearity. You can see that multicollinearity is not difficult to detect, but the exact nature of the multicollinearity is difficult to diagnose.

Combating the effects of multicollinearity is even more difficult. In fact, there is no universally optimum strategy for this task. Furthermore, many results obtained from the different methods are of questionable validity and usefulness.

The most obvious and most frequently used strategy is to create a model with fewer independent variables. Since there is often no a priori criterion for the selection of variables, an automated, data-driven search procedure is most frequently used. Use of such variable selection procedures is often called *model building*. Since these methods use the data to select the model, such methods are more appropriately called *data dredging*. This approach has two major drawbacks:

1. This approach is not appropriate if it is important to find the structure of the regression relationship for all variables in the model.
2. The results of the variable selection procedures often do not provide a clear choice of an optimum model, that is, the model with the minimum error mean square for a given set of variables, especially when there is multicollinearity.

Procedures for variable selection using the SAS System are presented in section **4.3**.

When it is important to study the nature of the regression relationship, it is sometimes useful to use multivariate techniques. Then you can study the structure of multicollinearity and consequently use the results of such a study to enhance the regression analysis. One such multivariate method is *principal components analysis*. This method generates a set of artificial, uncorrelated variables that can then be used in a regression model. Implementing a principal component regression using the SAS System is presented in section **4.4**.

Another method that you can use to study the structure of the regression relationship is that of biased estimation. The application of least squares is known to provide unbiased estimates of the parameters of a linear model. Although unbiased estimates are usually desired, situations exist where biased estimates may have smaller variances and, thus, provide useful estimates. The precision of biased and unbiased estimates can be compared by the mean squared error that is defined as the variance plus the square of the bias. Multicollinearity is a major contributor to large variances of estimates of regression coefficients; hence, a biasing of estimates that reduces the effect of multicollinearity may provide estimates with smaller mean squared error.

A number of biased estimation procedures have been proposed for regression coefficients in the presence of multicollinearity. The most popular of these is ridge regression (see Draper and Van Nostrand 1979), although other methods, especially those that use information from an eigenanalysis, may be useful.

Biased estimation methods appear to be a useful method for predicting the response for one sample using models estimated from another sample. One reason for this is that variable selection is quite unstable in the presence of multicollinearity; hence, using the results of a variable selection based on one sample may not provide a good model for another sample. Biased estimation can also be useful for variable selection. This method may delete variables because of weak relationships to the response variable rather than due to multicollinearity.

Ridge regression can be implemented in the SAS System with PROC RIDGEREG documented in the *SUGI Supplemental Library User's Guide*. Ridge regression can also be implemented by using the matrix formulation using PROC MATRIX or PROC IML. Because ridge regression and other biased estimation methods are quite controversial, and because these methods are not widely used, these procedures are not presented here.

4.2 DETECTING MULTICOLLINEARITY

Because multicollinearity is essentially a "messy data" problem, the detection and treatment of this problem are not adequately illustrated with small, compact examples of the type used in previous chapters. For this reason, a "messy" example was chosen for this chapter. It should be emphasized that not all multicollinearity examples show so little promise of a useful solution, but, on the other hand, many are much worse!

The example for this topic concerns the determination of the percentage of fat (FAT) in pork bellies. Since this measurement is determined by a rather expensive process, it is of interest to see if this value can be determined from other more easily measured properties of the pork carcass. Ten such variables have been measured on a sample of forty-five pork carcasses.* These variables are as follows:

AVBF	is an average of three measures of back fat thickness.
MUS	is a muscling score for the carcass. The higher the number is, the more muscle and less fat.
LEA	is loin eye area.
DEP	is an average of three measures of fat depth opposite the tenth rib.
LWT	is live weight of the carcass.
CWT	is weight of the slaughtered carcass.
WTWAT	is a measure used to determine specific gravity.
DPSL	is the average of three determinations of depth of the belly.
LESL	is the average measure of leanness of three cross sections of the belly.
BELWT	is total weight of the belly.

The data are given in **Output 4.1**.

* The original data set contained several additional variables; a few have been arbitrarily deleted to make the example more manageable.

Output 4.1 Bellyfat Data

AVBF	MUS	LEA	DEP	LWT	CWT	WTWAT	DPSL	LESL	BELWT	FAT
1.30	14	5.00	1.27	239	187	4.10	1.50000	10.0000	14.35	51.4
1.57	9	4.10	1.47	229	175	3.83	1.56667	6.0000	13.76	58.0
1.68	11	4.20	1.60	223	172	3.38	1.53333	5.0000	12.99	51.0
1.58	9	4.30	1.73	210	160	3.38	1.40000	8.6667	12.42	54.5
1.18	13	5.00	1.13	234	178	3.86	1.53333	8.0000	13.77	53.1
1.98	13	4.20	1.97	239	185	3.24	1.60000	4.3333	14.06	57.1
1.28	9	4.80	1.23	226	172	3.83	1.30000	5.6667	12.32	55.3
1.62	11	5.50	1.57	225	172	4.14	1.50000	9.3333	12.82	49.9
1.53	10	4.10	1.80	227	173	3.10	1.76667	4.6667	10.93	57.3
1.20	10	5.50	1.00	215	164	4.21	1.46667	12.0000	11.10	46.0
1.67	10	5.10	1.60	230	180	3.95	1.66667	7.6667	12.49	54.1
1.88	10	4.70	1.67	233	181	3.82	1.43333	5.3333	12.72	61.6
1.50	9	4.90	1.60	212	162	3.71	1.60000	6.6667	12.87	56.1
1.47	14	5.40	1.17	244	192	4.32	1.36667	8.3333	13.93	51.7
1.38	11	5.05	1.20	236	180	4.43	1.40000	12.0000	13.19	50.9
1.88	8	3.30	2.17	217	166	2.61	1.60000	4.0000	12.02	64.8
1.72	12	4.90	1.60	223	171	3.64	1.33333	8.3333	12.61	56.1
1.88	6	4.40	1.80	226	175	3.69	1.53333	8.0000	11.65	57.5
1.73	12	4.00	1.57	232	177	3.82	1.66667	5.6667	12.99	54.4
1.33	9	4.90	1.33	221	170	3.96	1.30000	7.0000	12.44	50.9
1.42	6	5.00	1.37	219	166	3.87	1.30000	10.0000	12.05	49.5
1.35	9	4.80	1.43	228	175	3.70	1.43333	10.0000	12.34	56.6
1.78	11	5.10	1.43	226	176	3.95	1.56667	9.6667	13.37	49.8
1.35	15	4.60	1.37	230	178	3.52	1.53333	7.0000	15.25	58.5
1.18	9	3.90	1.20	224	168	3.73	1.60000	7.3333	13.03	55.4
1.58	10	4.00	1.60	223	167	3.60	1.76667	7.0000	11.24	50.3
1.70	12	3.50	2.07	240	188	3.10	1.70000	5.0000	14.78	65.4
1.52	10	4.45	1.47	231	178	3.72	1.66667	6.3333	13.39	55.3
1.30	15	4.80	1.33	235	183	3.61	1.43333	6.6667	13.10	50.3
1.68	10	4.05	1.80	241	195	3.63	1.86667	6.6667	14.56	60.1
1.80	11	3.05	2.07	222	166	2.33	1.73333	3.6667	11.77	58.7
1.78	11	5.10	1.43	226	176	3.95	1.60000	7.0000	13.21	49.8
1.18	9	3.90	1.20	224	168	3.73	1.56667	5.6667	13.84	58.3
1.68	11	4.20	1.60	223	172	3.38	1.50000	9.6667	15.63	55.1
1.18	13	5.00	1.13	234	178	3.86	1.46667	8.0000	14.62	53.1
1.98	13	4.20	1.97	239	185	3.24	1.46667	6.3333	15.17	59.8
1.35	9	4.85	1.10	214	164	4.27	1.50000	9.6667	12.17	46.7
1.20	10	5.50	1.00	215	164	4.21	1.56667	12.6667	12.43	46.0
1.47	14	5.40	1.17	244	192	4.32	1.80000	7.6667	11.66	53.0
1.88	8	3.30	2.17	217	166	2.61	1.66667	4.6667	10.97	64.8
1.88	6	4.40	1.80	226	175	3.69	1.56667	6.6667	11.63	57.4
1.33	9	4.90	1.33	221	170	3.96	1.36667	7.0000	12.14	50.9
1.35	9	4.80	1.43	228	175	3.70	1.26667	6.3333	13.57	56.6
1.70	12	3.50	2.07	240	188	3.10	1.56667	5.3333	14.87	65.2
1.68	10	4.05	1.80	241	195	3.63	1.83333	6.3333	14.80	59.4

Three sets of statistics that are useful for determining the degree and nature of multicollinearity are available as MODEL statement options in PROC REG. These statistics are as follows:

1. the relative degree of significance of the model and its individual parameters
2. the variance inflation factors, often referred to as VIF
3. an analysis of the structure of the **X′X** matrix.

PROC REG is implemented with the following SAS statements:

```
PROC REG;
   MODEL FAT=AVBF MUS LEA DEP LWT CWT WTWAT DPSL
             LESL BELWT / VIF COLLINOINT;
```

The VIF option requests the calculation of the variance inflation factors. The VIF statistics are appended to the listing of parameter estimates of the output that is given in **Output 4.2**.

Output 4.2 Regression Using All Variables

```
DEP VARIABLE: FAT
                              ANALYSIS OF VARIANCE

                         SUM OF          MEAN
      SOURCE     DF     SQUARES        SQUARE      F VALUE      PROB>F

      MODEL      10   861.78956   86.17895635       13.437      0.0001
      ERROR      34   218.05844    6.41348343
      C TOTAL    44  1079.84800

         ROOT MSE      2.532486      R-SQUARE       0.7981
         DEP MEAN         55.06      ADJ R-SQ       0.7387
         C.V.          4.599502

                              PARAMETER ESTIMATES

                     PARAMETER        STANDARD       T FOR H0:
   VARIABLE  DF       ESTIMATE           ERROR     PARAMETER=0     PROB > |T|

   INTERCEP   1    24.85752840     21.01386546           1.183         0.2451
   AVBF       1    -5.47255074      3.65597016          -1.497         0.1437
   MUS        1    -0.58664327      0.30723799          -1.909         0.0647
   LEA        1    -0.98927490      1.70794636          -0.579         0.5663
   DEP        1     9.06495250      4.88651731           1.855         0.0723
   LWT        1     0.15048631      0.20471925           0.735         0.4673
   CWT        1     0.03036098      0.20963388           0.145         0.8857
   WTWAT      1    -1.81247934      2.96541371          -0.611         0.5451
   DPSL       1    -2.14189312      3.54748177          -0.604         0.5500
   LESL       1    -0.43270697      0.28992712          -1.492         0.1448
   BELWT      1     0.68668721      0.47218237           1.454         0.1550

                      VARIANCE
   VARIABLE  DF      INFLATION

   INTERCEP   1              0
   AVBF       1     5.39256048
   MUS        1     3.13858275
   LEA        1     7.95643967
   DEP        1    17.00146970
   LWT        1    22.71827368
   CWT        1    24.57954351
   WTWAT      1    12.43644447
   DPSL       1     1.88110915
   LESL       1     2.64865964
   BELWT      1     2.28670951
```

The COLLINOINT option (or COLLIN, see discussion in section **4.2.3**) provides for the analysis of structure that is given in **Output 4.3**.

Output 4.3 PROC REG with the COLLINOINT Option

```
                          CONDITION  VAR PROP  VAR PROP  VAR PROP  VAR PROP  VAR PROP
NUMBER  EIGENVALUE           NUMBER  AVBF      MUS       LEA       DEP       LWT

     1    4.040977         1.000000    0.0065    0.0000    0.0059    0.0031    0.0001
     2    2.967376         1.166963    0.0000    0.0221    0.0005    0.0001    0.0043
     3    0.877602         2.145826    0.0156    0.0204    0.0065    0.0000    0.0011
     4    0.758552         2.308079    0.0893    0.0231    0.0118    0.0071    0.0002
     5    0.497983         2.848628    0.0046    0.0147    0.0053    0.0011    0.0050
     6    0.479097         2.904234    0.0358    0.3889    0.0182    0.0014    0.0049
     7    0.226938         4.219774    0.0732    0.0000    0.0395    0.0105    0.0171
     8    0.096617         6.467208    0.2471    0.1047    0.5933    0.0801    0.0113
     9    0.034525        10.818673    0.5030    0.3534    0.1144    0.7708    0.0425
    10    0.020333        14.097542    0.0250    0.0727    0.2046    0.1258    0.9136
```

(continued on next page)

(continued from previous page)

NUMBER	VAR PROP CWT	VAR PROP WTWAT	VAR PROP DPSL	VAR PROP LESL	VAR PROP BELWT
1	0.0001	0.0037	0.0102	0.0154	0.0004
2	0.0038	0.0006	0.0015	0.0000	0.0264
3	0.0024	0.0096	0.2232	0.0110	0.1253
4	0.0012	0.0014	0.2387	0.0153	0.0034
5	0.0012	0.0000	0.1429	0.2887	0.3062
6	0.0048	0.0067	0.0085	0.0307	0.0422
7	0.0077	0.0512	0.0848	0.5760	0.2711
8	0.0157	0.1187	0.1071	0.0428	0.0753
9	0.0001	0.8029	0.0351	0.0199	0.0248
10	0.9629	0.0051	0.1480	0.0002	0.1249

4.2.1 Model and Parameters

The hypothesis tests statistics on the model and regression coefficients clearly indicate the existence of multicollinearity. The entire model is statistically significant, $p<0.0001$, yet the smallest p value for a regression coefficient is 0.0647, which is not significant at the 0.05 level. This result is a natural consequence of multicollinearity: the overall model may fit the data quite well, but because several independent variables are measuring similar physical phenomena, it is difficult to determine which variables are important in the regression relationship.

4.2.2 Variance Inflation Factors

The variance inflation factors are useful in determining which variables may be involved in the multicollinearities. For the ith coefficient, the variance inflation factor is defined as $1/(1-R_i^2)$, where R_i^2 is the coefficient of determination of the regression of the ith independent variable on all other independent variables. It can be shown that the variance of the estimate of that coefficient is larger by that factor than it would be if there were no multicollinearity, that is, if the independent variables were uncorrelated. In other words, the existence of multicollinearity has increased the instability of the coefficient estimate. There are no formal criteria for determining the magnitude of variance inflation factors that cause poorly estimated coefficients. Some authorities have stated that values exceeding five or ten should be cause for concern, but these limits are arbitrary. Actually, if the coefficient of determination of the regression with the dependent variable is low, coefficient estimates with relatively small variance inflation factors may still create unstable coefficient estimates. In **Output 4.2**, the regression R-SQUARE value is 0.7981. Since $1/(1-R^2)=4.953$, you can see that any variables associated with VIF values exceeding 4.953 are more closely related to the other independent variables than they are to the dependent variable. Six coefficients show VIF values exceeding this value, and of these, four exceed the arbitrary value of ten. Thus, it is reasonable to assume that at least six coefficient estimates suffer from the effects of multicollinearity. You may therefore conclude that these coefficients may be statistically nonsignificant due to multicollinearity and not due to the fact that they are not related to the dependent variable.

4.2.3 Analysis of Structure

An eigenanalysis of matrices derived from the sums of squares and cross products of these variables yields one of the many analyses of the structure of relationships among a set of variables. Such an eigenanalysis can be performed in two ways:

1. Use the raw (not centered) variables including the dummy variable to estimate the intercept and scale the variables such that the **X′X** matrix has ones along the diagonal. This method is implemented with the COLLIN option in PROC REG.
2. Use scaled and centered variables and exclude the dummy variable. In this case **X′X** is the correlation matrix; this method is implemented with the COLLINOINT option in PROC REG.

The first method arises from the use of the dummy variable to estimate the intercept. This makes it tempting to think of that coefficient simply as a coefficient for a variable that can be involved in multicollinearity. This is not always an appropriate conclusion. The intercept is an estimate of the response at the origin, that is, where all independent variables are zero. For most applications the intercept is an arbitrarily chosen point, and the intercept represents an extrapolation far beyond the reach of the data. The intercept is chosen for mathematical convenience rather than to provide a useful parameter. For this reason, the inclusion of the intercept in the study of multicollinearity can be useful only if the intercept has some physical interpretation that is within reach of the actual data space.

Centering the variables places the intercept at the means of all the variables. In this case, the intercept has no effect on the multicollinearity of the other variables. (For further information, see Belsley et al. 1984.) Centering is also consistent with the computation of the variance inflation factors that are based on first centering the variables. Since the origin cannot exist in this example, the COLLINOINT option is specified.

The eigenanalysis implemented by the COLLINOINT option provides the eigenvalues and variance proportions associated with the eigenvalues. A related analysis, principal components analysis, is presented in section **4.4**.

The portion of the output from the COLLINOINT option is shown in **Output 4.3**. The first column of the output consists of the eigenvalues of the correlation matrix of the set of independent variables arranged from largest to smallest. The severity of multicollinearity is revealed by the relative magnitudes of these eigenvalues. Note that since eigenvalues of zero indicate linear dependencies or exact collinearities, small eigenvalues indicate near linear dependencies or high degrees of multicollinearity. There are two eigenvalues that may be considered small in this set, thus implying the possibility of two sets of very strong relationships among the variables. The square root of the ratio of the largest to smallest eigenvalue is called the *condition index*. This value provides a single statistic for indicating the severity of multicollinearity. The condition index is the last entry in the column identified by CONDITION NUMBER. The condition index for this example, 14.10, indicates some multicollinearity, although it does not reach 30, which is sometimes proposed as the value required to indicate a high degree of multicollinearity.

The output also includes the square root of the ratio of the largest eigenvalue to each of the other eigenvalues; these make up the other entries in that column. The number of large values in this column is also an indicator of the number of near linear dependencies among the variables.

The portion of the output identifying the variance proportions, VAR PROP, can be used to indicate the variables involved in the near linear dependencies. Relatively large values in any row corresponding to the small eigenvalues point to such variables. In this example, in the row for eigenvalue 9 (0.035), the variance proportions are largest for the variables DEP and WTWAT, thus implying a strong relationship involving these two variables. Likewise, the proportions in row 10 (0.020) point to a strong relationship involving LWT and CWT. These four variables also exhibit the largest variance inflation factors; hence, the variance proportions do not provide much additional information for this example.

4.3 VARIABLE SELECTION

If a number of variables in a regression analysis do not appear to contribute significantly to the predictive power of the model, you often try to find some suitable subset of important or useful variables. The use of regression methodology assumes that you specify the appropriate model, but there is often no intuitive way of knowing what this model is. Instead, it is customary to use an automated procedure to select a suitable subset of your data.

For any given number of independent variables, a variable selection procedure should provide the subset whose estimated equation produces the best fit, that is, the subset whose estimated equation produces the minimum residual sum of squares or, equivalently, the maximum coefficient of determination, R^2. Such a subset is called an *optimum subset.* Presently, examining all possible subsets is the only way to guarantee the finding of optimum subsets. This procedure requires the computation of 2^m regression equations for m independent variables for all subset sizes. Modern computing capabilities make such a procedure feasible, but only for models with a moderate number of variables. A procedure for computing all possible combinations is implemented in the SAS System by PROC RSQUARE (see section **4.3.2**). This is not usually recommended for models containing more than twenty variables.

Popular alternatives to the examination of all possible subsets are the *stepping* or *step-type* procedures that add and delete variables one at a time until, by some criterion, a reasonable stopping point has been reached. For the most part, these procedures are less likely to find optimum models if multicollinearity exists. A number of stepping procedures are implemented in the SAS System with PROC STEPWISE.

Section **4.3.1** begins by implementing several stepping procedures on the bellyfat data. PROC RSQUARE is then used on the same data. Finally, since these procedures often provide a bewildering array of results, methods to summarize these results are presented in section **4.3.3**.

4.3.1 PROC STEPWISE

PROC STEPWISE allows the implementation of one or more of five different step-type variable selection procedures.

1. **Forward selection (FORWARD)** begins by finding the variable that produces the optimum one-variable subset, that is, the variable with the largest R^2. In the second step, the procedure finds that variable which, when added to the already chosen variable, results in the largest reduction in the residual sum of squares (or increase in R^2). The third step finds that variable which, when added to the two already chosen, gives the minimum residual sum of squares (or largest increase in R^2). The process continues until no variable considered for addition to the model provides a reduction in sum of squares considered statistically significant at a level you specify (see SLE specification below). An important feature of this method is that once a variable has been selected, it stays in the model.

2. **Backward elimination (BACKWARD)** begins by computing the regression with all independent variables. The statistics for the partial coefficients are examined to find the coefficient that contributes least to the fit of the model, that is, the coefficient with the largest *p* value (or smallest partial *F* value). The corresponding variable is deleted from the model, and the resulting equation is examined for the variable now contributing the least. This variable is then deleted and the procedure is continued. The procedure stops when all coefficients remaining in the model are statistically significant at a level you specify (see SLS specification below). With this method, once a variable has been deleted, it may never reenter the model.
3. **Stepwise selection (STEPWISE)** begins like forward selection, but after a variable has been added to the model, the resulting equation is examined to see if any coefficient has a sufficiently large *p* value (specified by the user) that suggests that a backward elimination procedure should be implemented. This procedure continues until no additions or deletions are indicated according to significance levels you choose.
4. **Maximum R^2 improvement (MAXR)** begins by selecting one- and two-variable models, as in forward selection. At that point, the procedure examines all possible pairwise interchanges with the variables not in the model. The interchange resulting in the largest increase in R^2 is implemented. This process is repeated until no pairwise interchange improves the model. At this point, as in forward selection, a third variable is selected. Then the interchanging process is implemented again, and so forth. This method usually examines more models and requires more computer time than the other three, but this method also tends to have a better chance of finding more nearly optimum models. In addition, statistics generated from the larger number of models that are examined with this method may be of some value in the final evaluation process.
5. **Minimum R^2 improvement (MINR)** is similar to maximum R^2 improvement, except that interchanges are implemented for those variables with minimum improvement. Since interchanges are not implemented when R^2 is decreased, the final results are quite similar to those of the maximum R^2 improvement method, except that a larger number of equations are examined.

PROC STEPWISE is implemented as follows:

```
PROC STEPWISE;
   MODEL  dependent variables=independent variables / options;
```

The options specify which selection procedure(s) are to be implemented as well as the parameters for starting and stopping rules. The options for implementing the procedures are as follows:

F or FORWARD
: requests the forward selection.

B or BACKWARD
: requests the backward elimination.

STEPWISE
requests the stepwise procedure.

MAXR
requests the maximum R^2 procedure.

MINR
requests the minimum R^2 procedure.

If you do not specify a procedure option, the stepwise procedure is the default. The desired significance levels for stopping the forward selection, backward elimination, and stepwise procedures are implemented as follows:

SLE=*value* or SLENTRY=*value*
specifies the significance level for adding variables in the forward selection mode. If you do not specify the level, the default is 0.50 for forward selection and 0.15 for stepwise.

SLS=*value* or SLSTAY=*value*
specifies the significance level for the backward elimination mode. If you do not specify the level, the default is 0.10 for backward elimination and 0.15 for stepwise.

The smallest permissible value for SLS is 0.0001, which almost always ensures that the final equation obtained by backward elimination contains only one variable. The maximum SLE of 0.99 usually includes all variables when forward selection stops. It is important to note that because variable selection is an exploratory rather than confirmatory analysis, the SLE and SLS values do not have the usual connotation, that is, the probabilities of erroneously rejecting the null hypothesis of the nonexistence of the coefficients in any selected model.

MAXR and MINR selection do not use significance levels, but you can alter the starting and stopping of this procedure with the following options:

START=*s*
specifies that the procedure starts with the first *s* independent variables listed in the MODEL statement.

STOP=*s*
specifies that the procedure stops when the best *s*-variable model has been found.

Finally, some variables may be required to remain in the model. The following option provides for this:

INCLUDE=*n*
forces the first *n* independent variables in the MODEL statement to be ineligible for elimination. This option is available for all selection procedures.

An additional option, DETAILS, provides information on the statistics used for the selection process. This option is available for the FORWARD and STEPWISE options because it provides statistics on the variables not included in the model.

The use of PROC STEPWISE is illustrated by requesting backward elimination for the bellyfat data. The following SAS statements are needed to do this:

```
PROC STEPWISE;
   MODEL FAT=AVBF MUS LEA DEP LWT CWT WTWAT DPSL
            LESL BELWT / B SLS=.0001 DETAILS;
```

Here, SLSTAY is set at 0.0001 so that the entire selection process can be examined. The output for this procedure is quite voluminous; hence, only steps 1, 2, and 4 (for 9, 8, and 6 variables) and the final summary are shown. The results are presented in **Output 4.4, 4.5, 4.6.** The first portion describing the full model and the statistics for the first deletion are given in **Output 4.2.**

Output 4.4 Backward Elimination

```
STEP 0    ALL VARIABLES ENTERED        R SQUARE = 0.79806562
                                       C(P) =     11.00000000

                DF          SUM OF SQUARES       MEAN SQUARE            F        PROB>F

REGRESSION      10            861.78956350       86.17895635        13.44        0.0001
ERROR           34            218.05843650        6.41348343
TOTAL           44           1079.84800000

                     B VALUE        STD ERROR        TYPE II SS            F        PROB>F

INTERCEPT      24.85752840
AVBF           -5.47255074       3.65597016       14.37038105         2.24        0.1437
MUS            -0.58664327       0.30723799       23.38257478         3.65        0.0647
LEA            -0.98927490       1.70794636        2.15168903         0.34        0.5663
DEP             9.06495250       4.88651731       22.07121086         3.44        0.0723
LWT             0.15048631       0.20471925        3.46553734         0.54        0.4673
CWT             0.03036098       0.20963388        0.13452483         0.02        0.8857
WTWAT          -1.81247934       2.96541371        2.39590463         0.37        0.5451
DPSL           -2.14189312       3.54748177        2.33802538         0.36        0.5500
LESL           -0.43270697       0.28992712       14.28578375         2.23        0.1448
BELWT           0.68668721       0.47218237       13.56414325         2.11        0.1550

BOUNDS ON CONDITION NUMBER:     24.57954,        1000.398
------------------------------------------------------------------------------

                         STATISTICS FOR REMOVAL: STEP 1
                                  DF = 1,34

                                   PARTIAL       MODEL
                         VARIABLE     R**2        R**2

                         AVBF       0.0133      0.7848
                         MUS        0.0217      0.7764
                         LEA        0.0020      0.7961
                         DEP        0.0204      0.7776
                         LWT        0.0032      0.7949
                         CWT        0.0001      0.7979
                         WTWAT      0.0022      0.7958
                         DPSL       0.0022      0.7959
                         LESL       0.0132      0.7848
                         BELWT      0.0126      0.7855
```

The initial step, labeled STEP 0, is the result of implementing the full model. This initial step is a condensed version of the output from PROC REG (**Output 4.2**). One minor difference is that the Type II (partial) sums of squares and resulting *F* statistics are given for testing the significance of the coefficients. The *F* statistics are equal to the squares of the *t* statistics for the partial coefficients produced by PROC REG (section **2.4.2**) and produce the same *p* values.

The C(P) statistic, located on **Output 4.4** below the R SQUARE value, is used to measure the merit of a selected subset. The C(P) statistic, proposed by Mallows (1973), is a measure of total squared error for a subset model containing *p* independent variables. The total squared error is a measure of the error variance plus the bias introduced by failing to include important variables in a model. Therefore, the total squared error may indicate when variable selection is deleting too many variables. The C(P) statistic is computed with the following equation:

$$C(P) = (SSE(P)/MSE) - (N - 2P) + 1 \quad .$$

MSE is the error mean square for the full model (or some other estimate of pure error).

SSE(P) is the error sum of squares for the subset model containing p independent variables (NOT including the intercept).*

N is total sample size.

When C(P)>(p+1), there is evidence of bias due to an incompletely specified model. If there are values of C(P)<(p+1), the full model is said to be overspecified; that is, it probably contains too many variables. Also, for any given number of selected variables, larger C(P) values indicate equations with larger error mean squares.

Mallows recommends that C(P) be plotted against p, and he further recommends the model where C(P) first approaches (p+1) starting from the full model. A plot of this type is presented in section **4.3.3**.

In **Output 4.4**, the listing of coefficients and their statistics are followed by the BOUNDS ON CONDITION NUMBER. These statistics are useful for detecting possible round off errors due to multicollinearity. These numbers must be exceedingly large (in the millions) before round off error may become serious.

In step 0, you can see that the coefficient for CWT has the smallest Type II sum of squares (and, of course, the smallest F and largest p values). The next step is to delete that variable since it contributes least to the fit of the equation.

The bottom portion of the output, produced by the DETAILS option, essentially reproduces the statistics for the full model. These statistics provide the partial R^2 (PARTIAL R**2) values that indicate the loss in the model R^2 by dropping each variable. These statistics also provide the model R^2 values (MODEL R**2) that indicate the R^2 values that would be obtained by deleting that variable.

The statistics for the model obtained by deleting CWT are given in **Output 4.5**.

Output 4.5 Backward Elimination Procedure with Variable CWT Removed

```
STEP 1    VARIABLE CWT REMOVED            R SQUARE = 0.79794104
                                          C(P) =     9.02097531

                  DF          SUM OF SQUARES      MEAN SQUARE          F        PROB>F

REGRESSION         9          861.65503867       95.73944874       15.36        0.0001
ERROR             35          218.19296133        6.23408461
TOTAL             44         1079.84800000

                     B VALUE       STD ERROR       TYPE II SS          F        PROB>F

INTERCEPT      22.53372991
AVBF           -5.50625652      3.59716435      14.60713394       2.34        0.1348
MUS            -0.59501051      0.29750688      24.93606591       4.00        0.0533
LEA            -0.89860501      1.56668721       2.05090606       0.33        0.5699
DEP             9.27389624      4.60293138      25.30627109       4.06        0.0517
LWT             0.17800926      0.07505571      35.06633497       5.62        0.0233
WTWAT          -1.76451162      2.90535251       2.29945054       0.37        0.5476
DPSL           -1.94998663      3.24437022       2.25203507       0.36        0.5517
LESL           -0.43390876      0.28572632      14.37702474       2.31        0.1378
BELWT           0.71198007      0.43252009      16.89257309       2.71        0.1087

BOUNDS ON CONDITION NUMBER:      15.51951,       472.2044
------------------------------------------------------------------------------------
```

(continued on next page)

* In the original presentation of the C(P) statistic (Mallows 1973), the intercept coefficient is also considered as a candidate for selection; hence, the number of variables in the model is one more than what is defined here. As implied in the discussion of the COLLIN option, deleting the intercept is not usually useful.

(continued from previous page)

```
STATISTICS FOR REMOVAL: STEP 2
            DF = 1,35

                PARTIAL      MODEL
VARIABLE           R**2       R**2

AVBF             0.0135     0.7844
MUS              0.0231     0.7748
LEA              0.0019     0.7960
DEP              0.0234     0.7745
LWT              0.0325     0.7655
WTWAT            0.0021     0.7958
DPSL             0.0021     0.7959
LESL             0.0133     0.7846
BELWT            0.0156     0.7823
```

Here you see that the partial F values have not increased, so a high degree of multicollinearity is suspected. For this model you can see that the variable LEA has the smallest partial F value; therefore, this variable is the candidate for deletion in the next step. Note that two other variables (WTWAT and DPSL) have partial F values that are almost as small, indicating that the selection process could easily be altered by very small changes in the data.

The selection process continues with steps two and three. Only the summary statistics are presented here:

```
STEP 2    VARIABLE LEA REMOVED     R SQUARE = 0.79604179
                                   C(P) =     7.34075568

STEP 3    VARIABLE DPSL REMOVED    R SQUARE = 0.79515807
                                   C(P) =     5.48954809
```

Note that R^2 is decreased only slightly and C(P) values remain less than $p+1$ with these deletions. The statistics for the six-variable regression resulting from step 4 are shown in **Output 4.6**.

Output 4.6 Backward Elimination with Variable WTWAT Removed

```
STEP 4    VARIABLE WTWAT REMOVED      R SQUARE = 0.78781875
                                      C(P) =     4.72528113


                 DF        SUM OF SQUARES      MEAN SQUARE            F      PROB>F

REGRESSION        6          850.72450159     141.78741693        23.52      0.0001
ERROR            38          229.12349841       6.02956575
TOTAL            44         1079.84800000

                      B VALUE       STD ERROR      TYPE II SS            F      PROB>F

INTERCEPT        16.10461148
AVBF             -7.92198257      2.92430917     44.24938396         7.34      0.0101
MUS              -0.55034422      0.22980152     34.58191748         5.74      0.0217
DEP              13.19387520      2.60710596    154.42350816        25.61      0.0001
LWT               0.13245217      0.05875331     30.64358109         5.08      0.0300
LESL             -0.61409832      0.24034104     39.36461196         6.53      0.0147
BELWT             0.85301426      0.39015919     28.82139114         4.78      0.0350

BOUNDS ON CONDITION NUMBER:     5.147696,      97.63335
--------------------------------------------------------------------------------
```

(continued on next page)

(continued from previous page)

```
                    STATISTICS FOR REMOVAL: STEP 5
                              DF = 1,38

                                 PARTIAL     MODEL
                    VARIABLE        R**2      R**2

                    AVBF          0.0410    0.7468
                    MUS           0.0320    0.7558
                    DEP           0.1430    0.6448
                    LWT           0.0284    0.7594
                    LESL          0.0365    0.7514
                    BELWT         0.0267    0.7611

STEP 5    VARIABLE BELWT REMOVED        R SQUARE = 0.76112852
                                        C(P) =     7.21915586
STEP 6    VARIABLE MUS REMOVED          R SQUARE = 0.74245191
                                        C(P) =     8.36376488
STEP 7    VARIABLE LESL REMOVED         R SQUARE = 0.71488193
                                        C(P) =    11.00576535
STEP 8    VARIABLE LWT REMOVED          R SQUARE = 0.65468550
                                        C(P) =    19.14112962
STEP 9    VARIABLE AVBF REMOVED         R SQUARE = 0.59650199
                                        C(P) =    26.93757601
```

STEP	VARIABLE REMOVED	NUMBER IN	PARTIAL R**2	MODEL R**2	C(P)	F	PROB>F
1	CWT	9	0.0001	0.7979	9.0210	0.0210	0.8857
2	LEA	8	0.0019	0.7960	7.3408	0.3290	0.5699
3	DPSL	7	0.0009	0.7952	5.4895	0.1560	0.6952
4	WTWAT	6	0.0073	0.7878	4.7253	1.3257	0.2570
5	BELWT	5	0.0267	0.7611	7.2192	4.7800	0.0350
6	MUS	4	0.0187	0.7425	8.3638	3.0493	0.0886
7	LESL	3	0.0276	0.7149	11.0058	4.2819	0.0450
8	LWT	2	0.0602	0.6547	19.1411	8.6563	0.0053
9	AVBF	1	0.0582	0.5965	26.9376	7.0768	0.0110

These statistics provide information for step 5. You can see that in this model one variable (DEP) has a very large partial F, and all other variables have p values less than 0.05 (although these cannot be taken literally). This evidence suggests that the multicollinearity among these remaining variables may no longer be severe.

A summary of the backward elimination constitutes the final portion of the output. Of course, the numbers are the same that are found for the individual steps. Examination of the summary shows that the first several steps delete variables that are clearly not useful. The C(P) statistic drops rapidly below $(p+1)$ and starts to increase when the number of variables in the model drops below six. Therefore, it would appear that the six-variable model is the best one found by the backward elimination method.

Next, implement the forward selection procedure for the same data. The SAS statements are as follows:

```
PROC STEPWISE;
   MODEL FAT=AVBF MUS LEA DEP LWT CWT WTWAT DPSL LESL BELWT /
              F SLE=.9 DETAILS;
```

Since the output for the forward selection is also quite voluminous, **Output 4.7** and **4.8** show only the output for steps 1 and 4, which produce models with one and four variables, and the final summary.

If you specify the DETAILS option, each step starts with the statistics used to guide the selection process. The following statistics appear in the output for each variable not in the model:

TOLERANCE is an indicator of round off error; values near zero indicate possible difficulties. It is also the reciprocal of the variance inflation factor for that variable if it is added to the current model.

MODELR**2 is the R^2 for the model obtained by adding the variable.

F is the F ratio for testing the hypothesis that adding that variable does not contribute to the fit of the model.

PROB>F is the p value for that F ratio. The smaller the p value, the stronger the evidence for rejecting the null hypothesis.

Output 4.7 contains the output for step 1 that chooses the optimum one-variable model. In the first step, the above statistics are the statistics for each of the one-variable models. The output continues with the statistics describing the chosen model that uses the variable DEP, using the same format as used in the backward elimination.

Output 4.7 Forward Selection Procedure, Step 1

```
              FORWARD SELECTION PROCEDURE FOR DEPENDENT VARIABLE FAT

                         STATISTICS FOR ENTRY: STEP 1
                                  DF = 1,43

                                      MODEL
             VARIABLE     TOLERANCE    R**2            F        PROB>F

             AVBF                 1   0.2664     15.6190        0.0003
             MUS                  1   0.0078      0.3371        0.5646
             LEA                  1   0.5745     58.0660        0.0001
             DEP                  1   0.5965     63.5681        0.0001
             LWT                  1   0.0750      3.4873        0.0687
             CWT                  1   0.0780      3.6353        0.0633
             WTWAT                1   0.5218     46.9182        0.0001
             DPSL                 1   0.1008      4.8212        0.0336
             LESL                 1   0.4882     41.0200        0.0001
             BELWT                1   0.0635      2.9153        0.0950

STEP 1     VARIABLE DEP ENTERED            R SQUARE = 0.59650199
                                           C(P) =     26.93757601

                 DF         SUM OF SQUARES     MEAN SQUARE          F        PROB>F

REGRESSION        1           644.13148224    644.13148224      63.57        0.0001
ERROR            43           435.71651776     10.13294227
TOTAL            44          1079.84800000

                    B VALUE      STD ERROR        TYPE II SS          F        PROB>F

INTERCEPT       36.89653767
DEP             11.87671905     1.48962518      644.13148224      63.57        0.0001

BOUNDS ON CONDITION NUMBER:              1,              1
```

Step 2 chooses the variable LEA, and step 3 chooses LWT. The output for step 4 is given in **Output 4.8**.

Output 4.8 Forward Selection Procedure, Step 4

```
                         STATISTICS FOR ENTRY: STEP 4
                                 DF = 1,40

                                       MODEL
          VARIABLE     TOLERANCE        R**2          F          PROB>F

          AVBF          0.228864      0.7469        3.7269       0.0607
          MUS           0.586976      0.7348        1.7374       0.1950
          CWT           .0549062      0.7238        0.0718       0.7901
          WTWAT         0.160451      0.7258        0.3663       0.5485
          DPSL          0.729624      0.7395        2.4861       0.1227
          LESL          0.427177      0.7327        1.3973       0.2442
          BELWT         0.661399      0.7300        0.9852       0.3269

STEP 4    VARIABLE AVBF ENTERED           R SQUARE = 0.74690996
                                          C(P) =     7.61315621

               DF          SUM OF SQUARES      MEAN SQUARE          F        PROB>F

REGRESSION      4            806.54922892     201.63730723      29.51        0.0001
ERROR          40            273.29877108       6.83246928
TOTAL          44           1079.84800000

                B VALUE        STD ERROR      TYPE II SS             F        PROB>F

INTERCEPT   25.60903127
AVBF        -6.55736857       3.39670621     25.46366468          3.73        0.0607
LEA         -2.44503979       1.08674666     34.58540936          5.06        0.0300
DEP         12.09187806       3.44980006     83.94162176         12.29        0.0011
LWT          0.14123795       0.04479210     67.93251627          9.94        0.0031

BOUNDS ON CONDITION NUMBER:        7.954112,      65.47253
```

It is of interest to note that for that step, the smallest TOLERANCE value of 0.160 (for WTWAT) implies a maximum variance inflation factor of approximately 6, which implies a lesser degree of multicollinearity than in the full model.

Output 4.9 Forward Selection Procedure Summary

SUMMARY OF FORWARD SELECTION PROCEDURE FOR DEPENDENT VARIABLE FAT

STEP	VARIABLE ENTERED	NUMBER IN	PARTIAL R**2	MODEL R**2	C(P)	F	PROB>F
1	DEP	1	0.5965	0.5965	26.9376	63.5681	0.0001
2	LEA	2	0.0701	0.6666	17.1291	8.8360	0.0049
3	LWT	3	0.0567	0.7233	9.5835	8.4015	0.0060
4	AVBF	4	0.0236	0.7469	7.6132	3.7269	0.0607
5	MUS	5	0.0139	0.7608	7.2695	2.2698	0.1400
6	BELWT	6	0.0130	0.7738	7.0851	2.1795	0.1481
7	LESL	7	0.0187	0.7925	5.9444	3.3255	0.0763
8	WTWAT	8	0.0034	0.7959	7.3721	0.5994	0.4439
9	DPSL	9	0.0021	0.7979	9.0210	0.3612	0.5517
10	CWT	10	0.0001	0.7981	11.0000	0.0210	0.8857

The summary at the end (**Output 4.9**) is again similar to that for the backward elimination procedure. If you compare the results with those of the backward elimination, you can see that the methods produce somewhat different results. Although both methods choose the same one-variable equation, the forward method chooses better equations when selecting two-, three-, and four-variable models, while the backward elimination chooses better six- and seven-variable models. It must be emphasized that the pattern of differences in the results of the two methods may vary greatly with other data. Further comparisons are provided in section **4.3.3**.

The results obtained by implementing stepwise selection are not shown since the results are the same as the first four steps of the forward selection procedure. The results are the same when you use the default SLS and SLE values. Of course, other SLS and SLE values may provide different results. In general, stepwise selection is not often more effective than the forward or backward elimination methods.

Next, implement maximum R^2 selection. The SAS statements are as follows:

```
PROC STEPWISE;
   MODEL FAT=AVBF MUS LEA DEP LWT CWT WTWAT DPSL LESL BELWT /
            MAXR;
```

The search is not restricted by using a START= or STOP= option since you need to compare the results with those from the other procedures. Again, the entire output is not reproduced here since the appearance is the same as that for the backward elimination. Instead, a summary of the steps is provided in **Output 4.10.** Note again that the MAXR (as well as MINR) options do not implement a DETAILS option.

Output 4.10 Summary of Results of MAXR Selection

```
STEP 1     VARIABLE DEP ENTERED              R SQUARE = 0.59650199
                                             C(P) =     26.93757601
THE ABOVE MODEL IS THE BEST  1 VARIABLE MODEL FOUND.

STEP 2     VARIABLE LEA ENTERED              R SQUARE = 0.66663528
                                             C(P) =     17.12912784
THE ABOVE MODEL IS THE BEST  2 VARIABLE MODEL FOUND.

STEP 3     VARIABLE LWT ENTERED              R SQUARE = 0.72332918
                                             C(P) =      9.58348917
THE ABOVE MODEL IS THE BEST  3 VARIABLE MODEL FOUND.

STEP 4     VARIABLE AVBF ENTERED             R SQUARE = 0.74690996
                                             C(P) =      7.61315621
THE ABOVE MODEL IS THE BEST  4 VARIABLE MODEL FOUND.

STEP 5     VARIABLE MUS ENTERED              R SQUARE = 0.76082960
                                             C(P) =      7.26948485

STEP 5     LEA REPLACED BY LESL              R SQUARE = 0.76112852
                                             C(P) =      7.21915586
THE ABOVE MODEL IS THE BEST  5 VARIABLE MODEL FOUND.

STEP 6     VARIABLE BELWT ENTERED            R SQUARE = 0.78781875
                                             C(P) =      4.72528113
THE ABOVE MODEL IS THE BEST  6 VARIABLE MODEL FOUND.

STEP 7     VARIABLE WTWAT ENTERED            R SQUARE = 0.79515807
                                             C(P) =      5.48954809
THE ABOVE MODEL IS THE BEST  7 VARIABLE MODEL FOUND.

STEP 8     VARIABLE DPSL ENTERED             R SQUARE = 0.79604179
                                             C(P) =      7.34075568
THE ABOVE MODEL IS THE BEST  8 VARIABLE MODEL FOUND.

STEP 9     VARIABLE LEA ENTERED              R SQUARE = 0.79794104
                                             C(P) =      9.02097531
THE ABOVE MODEL IS THE BEST  9 VARIABLE MODEL FOUND.

STEP 10    VARIABLE CWT ENTERED              R SQUARE = 0.79806562
                                             C(P) =     11.00000000
THE ABOVE MODEL IS THE BEST 10 VARIABLE MODEL FOUND.
```

You can see that the maximum R^2 method selects the first four variables obtained by the forward selection method, but after selecting the fifth variable (MUS), it drops that variable in favor of LESL, with a small increase in R^2. The remainder of the selection produces the same subsets obtained by the backward elimination. You now see that three methods have given three different variable selections.

4.3.2 PROC RSQUARE

In order to compare the effectiveness of these selection methods, PROC RSQUARE is used. PROC RSQUARE guarantees finding the optimum subsets. The procedure is implemented in a manner similar to PROC STEPWISE. The invocation of the procedure is followed by a MODEL statement, which is followed by options that include the following:

START=*n*
: specifies the smallest number of regressors to be reported in the subset model.

STOP=*n*
: specifies the largest number of regressors to be reported in the subset model.

SELECT=*n*
: limits the printing of subsets to the *n* best since often a very large number of subsets are calculated.

CP
: specifies that the C(P) statistic be printed for all models.

Other options are available to print additional statistics that evaluate the selection process, as well as print the coefficients for selected models. This latter option must be used with care since in some situations it can become quite expensive and also produce a very large quantity of output. Finally, the results of the selection procedure may be output to a SAS data set. This example uses the following SAS statements:

```
PROC RSQUARE;
   MODEL FAT=AVBF MUS LEA DEP LWT CWT WTWAT DPSL LESL BELWT /
         CP SELECT=10;
```

No restriction is placed on the procedure since the goal is to compare the results with those of the other procedures. However, only the best ten equations are printed for each set; the default is ten if the number of independent variables is greater than ten. All selections are printed if the number is less than or equal to ten (as it is in this case). Furthermore, to save space, only that part of the output for one to five variables is reproduced, although the results of the entire procedure are used in the comparisons presented in section **4.3.3**. These results are shown in **Output 4.11**.

Output 4.11 All Possible Regressions

```
N=45        REGRESSION MODELS FOR DEPENDENT VARIABLE: FAT  MODEL: MODEL1

NUMBER IN   R-SQUARE        C(P)  VARIABLES IN MODEL
  MODEL

    1      0.00777770    126.062  MUS
    1      0.06349389    116.681  BELWT
    1      0.07501547    114.741  LWT
    1      0.07795212    114.247  CWT
    1      0.10081648    110.397  DPSL
    1      0.26644913  82.509080  AVBF
    1      0.48821709  45.169670  LESL
    1      0.52178760  39.517352  WTWAT
    1      0.57453537  30.636130  LEA
    1      0.59650199  26.937576  DEP
-------------------------------------
    2      0.62996562  23.303253  DEP CWT
    2      0.64290027  21.125427  DEP LWT
    2      0.64781689  20.297608  LEA LWT
    2      0.64794572  20.275917  WTWAT LWT
    2      0.64847911  20.186109  DEP BELWT
    2      0.65468550  19.141130  DEP AVBF
    2      0.65881273  18.446222  WTWAT CWT
    2      0.66159909  17.977078  DEP LESL
    2      0.66343046  17.668727  LEA CWT
    2      0.66663528  17.129128  DEP LEA
-----------------------------------------
    3      0.70232261  13.120397  DEP AVBF BELWT
    3      0.70342363  12.935017  LEA CWT WTWAT
    3      0.70631195  12.448705  DEP AVBF CWT
    3      0.70818762  12.132896  DEP LEA BELWT
    3      0.71019815  11.794379  DEP LESL BELWT
    3      0.71488193  11.005765  DEP AVBF LWT
    3      0.72255468   9.713892  DEP LEA CWT
    3      0.72332918   9.583489  DEP LEA LWT
    3      0.73026843   8.415116  WTWAT CWT MUS
    3      0.73556725   7.522946  WTWAT LWT MUS
---------------------------------------------
    4      0.74297612   8.275503  LEA LWT WTWAT MUS
    4      0.74325753   8.228122  DEP LEA CWT DPSL
    4      0.74364675   8.162589  WTWAT CWT MUS LESL
    4      0.74420968   8.067807  WTWAT LWT MUS LESL
    4      0.74554035   7.843760  LEA CWT WTWAT MUS
    4      0.74574756   7.808871  DEP LESL BELWT AVBF
    4      0.74652169   7.678530  DEP LEA CWT AVBF
    4      0.74690996   7.613156  DEP LEA LWT AVBF
    4      0.74989968   7.109772  WTWAT CWT MUS BELWT
    4      0.75896457   5.583504  WTWAT LWT MUS BELWT
---------------------------------------------------
    5      0.75911312   7.558493  WTWAT LWT MUS BELWT AVBF
    5      0.75944107   7.503274  DEP LESL BELWT AVBF MUS
    5      0.75968819   7.461666  WTWAT CWT MUS BELWT LWT
    5      0.76082960   7.269485  DEP LEA LWT AVBF MUS
    5      0.76112852   7.219156  DEP LESL AVBF LWT MUS
    5      0.76185504   7.096831  LEA LWT WTWAT MUS BELWT
    5      0.76203109   7.067190  DEP LEA CWT AVBF DPSL
    5      0.76638302   6.334448  DEP BELWT WTWAT LWT MUS
    5      0.76891677   5.907837  WTWAT CWT MUS BELWT LESL
    5      0.77283590   5.247966  WTWAT LWT MUS BELWT LESL
---------------------------------------------------------
```

The results indicate that none of the stepping procedures found all of the optimum subsets either independently or jointly.

4.3.3 Comparing Selection Methods

Plotting the C(P) values against the subset size is an instructive method for comparing the effectiveness of the different selection procedures, as well as for picking the most appropriate equation to use. At the present time, these C(P) values are available directly as a SAS data set only with PROC RSQUARE; hence, these values are manually transcribed from the computer output.* The resulting data set is given in **Output 4.12**.

Output 4.12 Summary of C(P) Values

CP PLOT

OBS	P	B	F	MAXR	BEST	SECOND	PP
1	1	26.94	26.94	26.94	26.94	30.64	2
2	2	19.14	17.13	17.13	17.13	17.67	3
3	3	11.01	9.58	9.58	7.52	8.42	4
4	4	8.36	7.61	7.61	5.58	7.11	5
5	5	7.22	7.27	7.22	5.25	5.91	6
6	6	4.73	7.09	4.73	4.73	5.92	7
7	7	5.49	5.94	5.49	5.49	5.94	8
8	8	7.34	7.37	7.34	7.34	7.37	9
9	9	9.02	9.02	9.02	9.02	9.33	10
10	10	11.00	11.00	11.00	11.00	11.00	11

In this table, the column labeled P indicates the number of variables selected; the columns labeled F, B, and MAXR give the C(P) values for the backward, forward, and maximum R^2 improvement methods, and the columns labeled BEST and SECOND give the C(P) values for the best (optimum) and second best equations found by the RSQUARE procedure. The variable PP represents the values of $p+1$, which are used as reference points for the C(P) statistic. The differences in the results of the procedures are evident.

These values are presented on two plots. One plot compares the results of the different procedures, and the other plot compares the best and second best equations found by the RSQUARE procedure. The following SAS statements are required:

```
PROC PLOT;
   PLOT B*P='B' F*P='F' MAXR*P='M' PP*P='*' BEST*P='1' / OVERLAY
        HAXIS=1 TO 10 BY 1 VAXIS=3 TO 10 BY 1;
   PLOT BEST*P='1' SECOND*P='2' PP*P='*' / OVERLAY
        HAXIS=1 TO 10 BY 1 VAXIS=3 TO 10 BY 1;
```

These plots are shown in **Output 4.13** and **4.14**. In order to provide a suitable scale for evaluating the results, the vertical scale on the plot has been specified to have a maximum value of 10, thus the C(P) values for $p=1$ and $p=2$ are out of range. The plotting symbols are identified in the table heading of **Output 4.12**.

* Currently, PROC PRINTTO is available on IBM systems to place SAS System output into a system file. It is, however, not worthwhile for the small amount of the output required here.

Output 4.13 C(P) Plot for Comparing Selection Methods

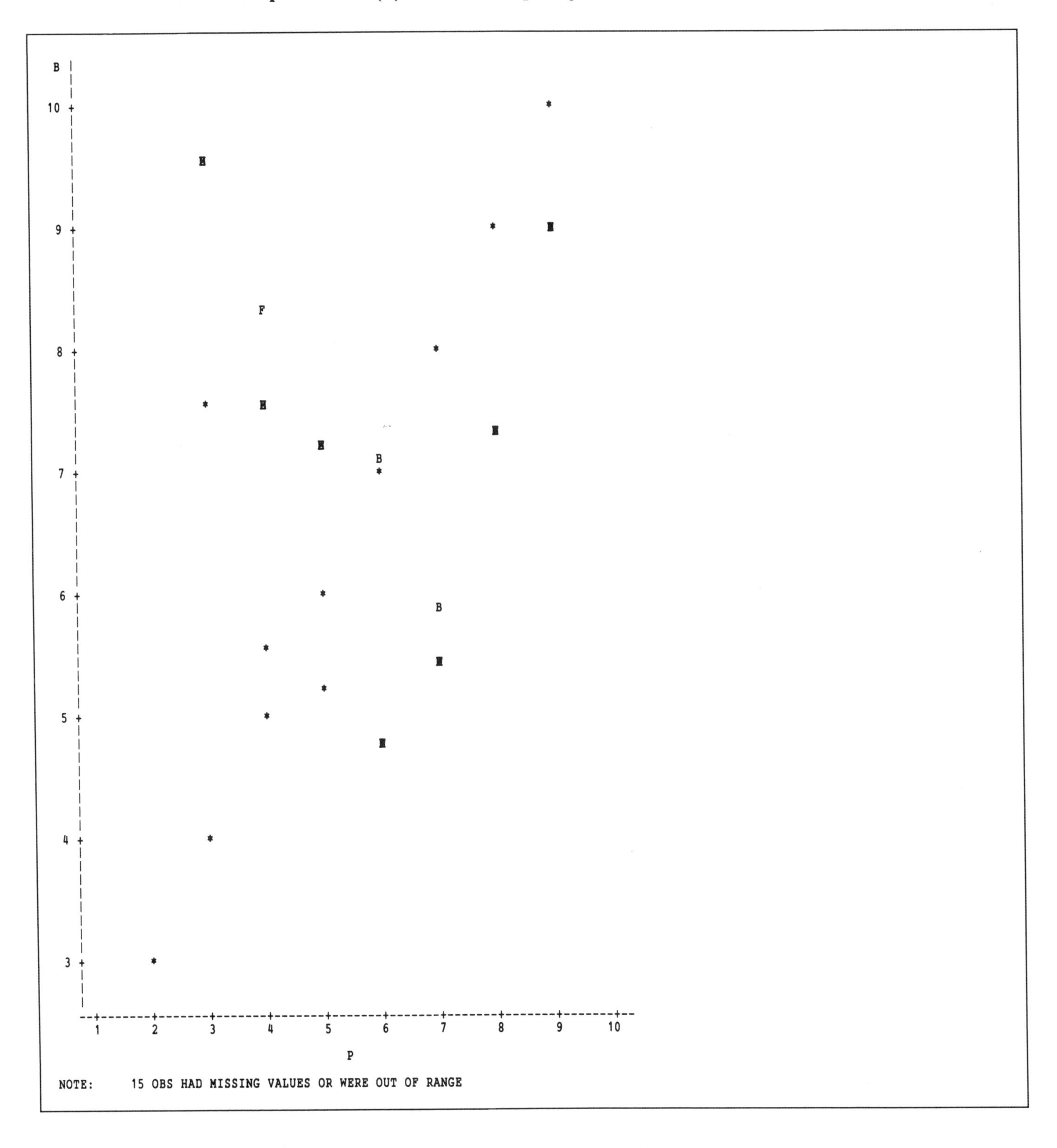

Output 4.14 C(P) Plot for Best and Second Best Models

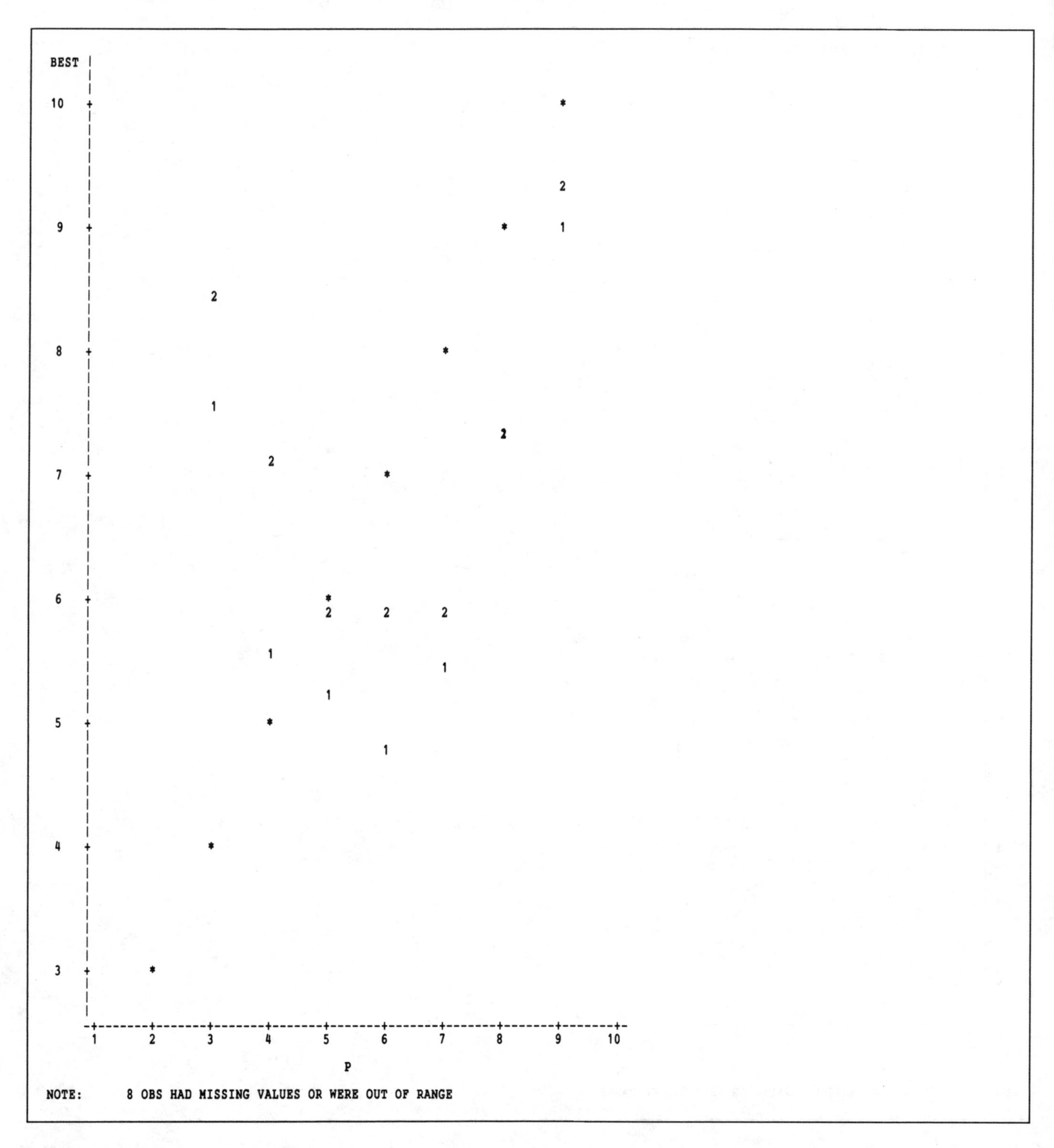

Output 4.13 clearly shows the differences in the performance of the stepping procedures. The results are a mixed bag; in this example the maximum R^2 procedure is somewhat superior to the other step procedures, but this will not always

happen. And, as already noted, none of the methods found all the optimum subsets. In general, contradictory results from these procedures are the rule rather than the exception, especially when models contain many variables. Also, none of the stepping procedures is clearly superior. Of course, the maximum R^2 procedure examines a larger number of subsets, which is of some advantage. Experience has shown that the backward elimination tends to perform somewhat better than forward selection. Of course, PROC RSQUARE does guarantee optimum subsets, but as previously noted, may become prohibitively expensive.

In **Output 4.14** you can see that as you drop from ten down to eight variables, there is little difference in C(P) values between the best and second best subsets. This is an indication of multicollinearity: it is possible to swap variables without much effect on the fit of the equation. As you proceed below eight variables, swapping variables has more effect; this effect is quite marked below five variables. Quite often, subset sizes showing a wider divergence between the C(P) values of the best and second best equations mark useful subset sizes. This criterion indicates that a four-variable subset may provide a useful model.

Generally, the most useful subset sizes are considered to be those between the minimum value of C(P) and the point where the values exceed $(p+1)$. This criterion suggests that the four- or five-variable models are appropriate. The optimum four-variable model is analyzed with PROC REG; the results are given in **Output 4.15.**

Output 4.15 Selected Regression

```
DEP VARIABLE: FAT
                              ANALYSIS OF VARIANCE

                           SUM OF          MEAN
       SOURCE     DF      SQUARES        SQUARE       F VALUE      PROB>F

       MODEL       4    819.56637     204.89159        31.488      0.0001
       ERROR      40    260.28163    6.50704076
       C TOTAL    44   1079.84800

           ROOT MSE       2.55089      R-SQUARE        0.7590
           DEP MEAN         55.06      ADJ R-SQ        0.7349
           C.V.          4.632928

                              PARAMETER ESTIMATES

                      PARAMETER        STANDARD        T FOR H0:
   VARIABLE  DF        ESTIMATE           ERROR      PARAMETER=0      PROB > |T|

   INTERCEP   1     20.41097172     11.36110854            1.797          0.0800
   WTWAT      1     -8.06179733      0.85584230           -9.420          0.0001
   LWT        1      0.28169114      0.05854602            4.811          0.0001
   MUS        1     -0.96743430      0.22925846           -4.220          0.0001
   BELWT      1      0.78634595      0.39906267            1.970          0.0557

                       VARIANCE
   VARIABLE  DF       INFLATION

   INTERCEP   1               0
   WTWAT      1      1.02099496
   LWT        1      1.83131543
   MUS        1      1.72244339
   BELWT      1      1.60984521
```

The results clearly show that the multicollinearity has been reduced. And, although the p values may not be taken literally, three of the coefficients do appear important. However, the inclusion of the BELWT variable is open to some question. Remember, however, that these results apply only to this particular data set; there is no guarantee that this subset model is optimal for the population.

4.4 PRINCIPAL COMPONENT REGRESSION

For some applications, redefinition of variables can be used to reduce multicollinearity. For example, if x and y exhibit a strong positive correlation, the variables $(x+y)$ and $(x-y)$ will exhibit a low correlation. In this example, LWT and CWT are both measures of the size of the carcass and are positively correlated. However, (LWT−CWT), which is a measure of the unusable portion of the carcass, will exhibit a low correlation with CWT. Such redefinitions can be quite useful in reducing multicollinearity but are not always applicable. Data-based redefinition of variables can be accomplished by multivariate analysis methods.

Principal components analysis is a multivariate analysis technique that attempts to describe interrelationships among a set of variables. Starting with a set of observed values on a set of m variables, this method uses a set of linear transformations to create a new set of variables called the *principal components*. The principal components have the following properties:

- The principal component variables are jointly uncorrelated.
- The first principal component has the largest variance of any linear function of the original variables (subject to a scale constraint). The second component has the second largest variance, and so forth.

Principal components are obtained by computing the eigenvalues and eigenvectors of the correlation or covariance matrix. In most applications the correlation matrix is used so that the components are not affected by the scales of measurement of the original variables.

The eigenvalues are the variances of the component variables. If the correlation matrix has been used, the sum of the variances of both the original and the component variables is equal to the number of variables. A set of eigenvalues of relatively equal magnitudes indicates that there is little multicollinearity, while a wide variation in magnitudes indicates severe multicollinearity. In fact, the number of large (usually greater than unity) eigenvalues may be taken as a very crude indication of the true number of variables (sometimes called factors) needed to describe the behavior of the full set of variables.

The eigenvectors are the coefficients for the linear transformation of the standardized variables (mean zero, variance one). These coefficients are used to create observed values of the new component variables and can be used to interpret the structure of these variables. The eigenvectors are also used to compute the correlations between the originally observed variables and the new factor level variables. These correlations, called *factor loadings* or *factor structures*, are usually better indicators of the nature of the component variables than the eigenvectors themselves since these correlations take into account the actual data values in the problem being studied. For more complete descriptions of principal components see Johnson and Wichern (1982) and Morrison (1976).

The set of principal component variables may then be used as independent variables in a regression analysis. Such an analysis is called principal component regression. Since the components are uncorrelated, there is no multicollinearity in the regression. Therefore, you can determine the important coefficients without resorting to variable selection techniques. Then, if the coefficients of the principal components transformations imply meaningful interpretation of the component variables, the regression may shed light on the underlying regression relationships. Unfortunately, such interpretations are not always clear-cut or obvious.

Principal component regression is illustrated on the bellyfat data. The analysis uses the following SAS System procedures:

- PROC FACTOR performs the principal component analysis.
- PROC REG performs the regression of the dependent variable on the set of component variables.

Each part is discussed separately, although normally both parts are performed sequentially in one job. The SAS statements for the principal component analysis are as follows:

```
PROC FACTOR OUT=PRIN N=10;
   VAR AVBF MUS LEA DEP LWT CWT WTWAT DPSL LESL BELWT;
```

PROC FACTOR provides a large number of different factor analysis procedures for analyzing the structure of a set of variables. The simplest, and default, method is that of principal components. The following additional PROC options are used here:

OUT=PRIN   creates a data set called PRIN that contains the results of the procedure. This is required for the subsequent implementation of PROC REG.

N=10   instructs the procedure to compute all ten principal components. Less than this number may be specified, but the full number is usually included for principal component regression.

The output from PROC FACTOR is given in **Output 4.16**.

Output 4.16 Principal Components Analysis

```
INITIAL FACTOR METHOD: PRINCIPAL COMPONENTS

                  PRIOR COMMUNALITY ESTIMATES: ONE

EIGENVALUES OF THE CORRELATION MATRIX:  TOTAL = 10        AVERAGE =    1

                       1         2         3         4         5
         EIGENVALUE  4.040977  2.967376  0.877602  0.758552  0.497983
         DIFFERENCE  1.073601  2.089774  0.119050  0.260568  0.018887
         PROPORTION    0.4041    0.2967    0.0878    0.0759    0.0498
         CUMULATIVE    0.4041    0.7008    0.7886    0.8645    0.9142

                       6         7         8         9        10
         EIGENVALUE  0.479097  0.226938  0.096617  0.034525  0.020333
         DIFFERENCE  0.252158  0.130322  0.062091  0.014192
         PROPORTION    0.0479    0.0227    0.0097    0.0035    0.0020
         CUMULATIVE    0.9622    0.9849    0.9945    0.9980    1.0000

              10 FACTORS WILL BE RETAINED BY THE NFACTOR CRITERION
```

(continued on next page)

(continued from previous page)

FACTOR PATTERN

	FACTOR1	FACTOR2	FACTOR3	FACTOR4	FACTOR5
AVBF	0.75551	-0.03027	0.25473	0.52624	0.07825
MUS	-0.04572	0.78151	-0.22210	-0.20428	-0.10703
LEA	-0.87598	0.18172	0.20013	0.23219	-0.10194
DEP	0.93117	-0.12092	0.02192	0.26369	0.06771
LWT	0.20051	0.92333	0.13643	0.05625	-0.16740
CWT	0.21716	0.91222	0.21455	0.12998	-0.08653
WTWAT	-0.86839	0.25781	0.30281	0.10064	0.00139
DPSL	0.55907	0.15840	0.56864	-0.50832	0.25818
LESL	-0.81623	0.00253	0.14965	0.15286	0.43547
BELWT	0.11591	0.72925	-0.46986	0.06711	0.41669

	FACTOR6	FACTOR7	FACTOR8	FACTOR9	FACTOR10
AVBF	0.21049	0.14254	-0.11152	-0.05686	-0.00747
MUS	0.52930	-0.00100	-0.05538	0.03636	-0.00971
LEA	0.18235	0.12728	0.20992	-0.03294	0.02594
DEP	0.07266	-0.09602	0.11272	0.12498	0.02974
LWT	-0.15923	-0.14133	-0.04895	-0.03393	0.09263
CWT	-0.16487	-0.09844	0.06011	0.00173	-0.09892
WTWAT	-0.13814	0.18115	-0.11741	0.10910	0.00514
DPSL	0.06050	0.09065	0.04338	-0.00887	0.01073
LESL	0.13666	-0.28032	-0.03253	-0.00792	-0.00042
BELWT	-0.14883	0.17868	0.04009	-0.00822	0.01087

The first section provides information on the eigenvalues of the correlation matrix. The eigenvalues are identified by column headings. Since there are ten variables, there are ten eigenvalues that are arranged in order from high to low.

The first entry in each column is the eigenvalue itself, which is the same obtained by the COLLINOINT option in PROC REG (**Output 4.3**).

Since the principal components are (by default) computed from the standardized variables, the sum of the variances for the ten variables is ten, and this value is a measure of the total variation inherent in the ten variables. Dividing each of the eigenvalues by this number gives the proportion of total variation accounted for by each of the components; these values are given in the row labeled PROPORTION. The last row gives the CUMULATIVE proportions. For each component, these cumulative proportions indicate the proportion of the total variation of the original set of variables explained by all components up to and including that component. For example, 78.86% of the total variation measured by the original ten variables is explained by only three components. Likewise, over 90% of the variation is explained by five components. In other words, a relatively small number of component variables can describe most of the variation of all ten variables. This is another indication that the original set of variables contains redundant information.

The second portion of the output (FACTOR PATTERN) contains the factor loadings. The original variables are identified by row labels, and the column labels identify the ten component variables (FACTOR1- FACTOR10). Large values of correlations for any factor are used to help identify or interpret the new variables, especially those associated with the large eigenvalues. The first component has large positive correlations with AVBF, DEP, DPSL, and LESL, which can be considered measures of overall fatness. The first component has negative correlations with LEA and WTWAT, which are measures of overall leanness. Thus, the first component may be interpreted as a measure of the fatness of the carcass. The second component exhibits high positive correlations with variables that can be considered indicators of the overall size of the carcass. None of the other components has high correlations with the original variables, but these other components may have some meaning to someone more familiar with this type of data.

It is not uncommon for a principal components analysis to produce results that are difficult to interpret. For this reason, rotation methods attempt to provide for more readily interpreted factor variables. Such rotations, as well as other methods of factor extraction, are available in PROC FACTOR, but these methods are beyond the scope and purpose of this book.

The remainder of the output from PROC FACTOR is of little interest for present purposes and is not presented here.

The implementation of the OUT=PRIN option in PROC FACTOR creates a data set called PRIN that contains all the variables in the original bellyfat data set as well as the ten new principal component variables, FACTOR1-FACTOR10. PROC REG is now used to perform the regression:

```
PROC REG;
   MODEL FAT=FACTOR1-FACTOR10 / SS2;
```

The output is shown in **Output 4.17**.

Output 4.17 Principal Component Regression

```
DEP VARIABLE: FAT
                              ANALYSIS OF VARIANCE

                          SUM OF          MEAN
      SOURCE     DF      SQUARES        SQUARE          F VALUE        PROB>F

      MODEL      10    861.78956   86.17895635           13.437        0.0001
      ERROR      34    218.05844    6.41348343
      C TOTAL    44   1079.84800

          ROOT MSE      2.532486       R-SQUARE          0.7981
          DEP MEAN         55.06       ADJ R-SQ          0.7387
          C.V.          4.599502

                              PARAMETER ESTIMATES

                      PARAMETER        STANDARD       T FOR H0:
   VARIABLE  DF        ESTIMATE           ERROR     PARAMETER=0        PROB > |T|

   INTERCEP   1     55.06000000      0.37752067         145.846            0.0001
   FACTOR1    1      4.04195706      0.38178657          10.587            0.0001
   FACTOR2    1      0.39727886      0.38178657           1.041            0.3054
   FACTOR3    1     -0.83295273      0.38178657          -2.182            0.0361
   FACTOR4    1      0.29402287      0.38178657           0.770            0.4465
   FACTOR5    1     -0.08867897      0.38178657          -0.232            0.8177
   FACTOR6    1     -1.27980607      0.38178657          -3.352            0.0020
   FACTOR7    1     -0.53147060      0.38178657          -1.392            0.1729
   FACTOR8    1      0.51549568      0.38178657           1.350            0.1859
   FACTOR9    1      0.28259319      0.38178657           0.740            0.4643
   FACTOR10   1      0.19182312      0.38178657           0.502            0.6186

   VARIABLE  DF     TYPE II SS

   INTERCEP   1      136422.16
   FACTOR1    1      718.84634
   FACTOR2    1     6.94454155
   FACTOR3    1    30.52765121
   FACTOR4    1     3.80377564
   FACTOR5    1     0.34601419
   FACTOR6    1    72.06775694
   FACTOR7    1    12.42828378
   FACTOR8    1    11.69237510
   FACTOR9    1     3.51379206
   FACTOR10   1     1.61902888
```

The statistics for the overall regression are, by definition, the same as for the original regression with ten variables (**Output 4.2**). However, the coefficients tell a different story. Since the component variables are standardized and uncorrelated, all coefficient estimates have equal standard errors: $\sqrt{s^2/(n-1)}$. Also, the

Type II sums of squares add to the regression sum of squares. The first component coefficient is the only really important coefficient. The coefficients for components three and six are marginally important.*

The most important component in the regression is the one identified with overall fatness. This component was most highly correlated with the variable DEP (r=0.93), which is indeed the variable chosen for the best single-variable model. However, the regression sum of squares for that one-variable model is 664.1, while the regression for the first component is 718.8; hence, the variable DEP does not adequately explain that component or explain the variation in bellyfat. Another interesting feature of the regression is the relatively large contribution (with a negative coefficient) of component 6. This component has a reasonably large correlation with only the muscling score. This is probably why MUS is one of the variables included in the optimum four-variable model.

In summary, principal component regression helped somewhat to interpret the structure of the regression. Although the interpretation is not overly clear-cut, this result is not uncommon with this type of analysis.

4.5 COMMENTS

In this chapter you have seen how several SAS System procedures can be used to estimate regression relationships when too many independent variables have been specified for the model. The results have not been entirely satisfactory, particularly when the different methods do not give the same results. This is partially due to the example that was purposely chosen to show that there is no unique solution to the problems created by multicollinearity. The example also shows that variable selection is not the panacea many users have been led to expect. This is particularly true of those models constructed through stepping procedures.

The ten variables used in the bellyfat example are only a subjectively chosen subset from the original data set; using a larger number of variables presents an even more confusing set of results. Furthermore, the forty-five observations used here are a randomly chosen half from a data set containing ninety-six observations. Variable selection and other methods do not give the same results when applied to the other fifty-one observations! Not all data sets produce contradictory results from different model building procedures, but even when they do, you should interpret the results with caution.

Again, it is important to stress that automated variable selection should not be used when prior information on the model can be used to choose a set of suitable variables. In other words, the brute force of the computer is no substitute for knowledge of your data and subject matter. For example, the relative cost of measuring the different independent variables should have a bearing on which variables to keep in a model whose purpose is to predict values of the dependent variable. Another case in point is the polynomial model where a natural ordering of parameters exists that suggests a predetermined order of variable selection (see Chapter 5).

Finally, the alternatives to variable selection, such as principal component regression or biased estimation, should be considered, especially if the primary objective is to study the structure of the regression relationship rather than to predict values of the dependent variable.

* You cannot use the p values literally, but the relative magnitude of the t statistics can be used to indicate relative importance of the coefficients.

5. Polynomial Models

5.1 INTRODUCTION

In the regression models discussed in earlier chapters, all relationships among variables have been described by straight lines. In this chapter, linear regression methods are used to estimate model parameters that cannot be described by straight lines. The most popular type of model used for this purpose is the *polynomial* model, in which the dependent variable is related to functions of the powers of one or more independent variables.

5.2 POLYNOMIAL MODELS WITH ONE INDEPENDENT VARIABLE

A one-variable polynomial model is written as follows:

$$y = \beta_0 + \beta_1 x + \beta_2 x^2 + \ldots + \beta_m x^m + \varepsilon$$

where y represents the dependent variable and x the independent variable. The highest exponent, or power, of x used in the model is known as the *degree* of the model. It is customary for a model of degree m to include all terms with lower powers of the independent variable.

A regression analysis for a polynomial model is produced using the values of all the required powers of the independent variable as the set of independent variables in a multiple linear regression model. Because the resulting model is linear in the parameters, all statistics and estimates produced by the implementation of that model have the same connotation as in any linear regression analysis, although the practical implications of some of the results may differ.

A special feature of polynomial regression is that the degree of the polynomial required to fit a set of data is not usually known a priori; hence, it is customary to build an appropriate polynomial model by sequentially fitting equations with higher order terms until a satisfactory degree of fit has been accomplished. In other words, you start by fitting a simple linear regression of y on x. Then you specify a model with linear and quadratic terms and decide if by adding the quadratic term you improve the fit by significantly reducing the residual mean square. You can then continue by adding and testing the contribution of a cubic term, then a fourth power term, and so forth, until no additional terms appear to be needed.

A polynomial regression model is illustrated here using data relating growth patterns of fish to water temperature. A curve describing how an organism grows with time is called a *growth curve.* This type of curve usually shows rapid initial growth that gradually becomes slower and may eventually stop altogether. Mathematical biologists have developed many sophisticated models to fit growth curves. A polynomial model often provides a convenient and easy approximation to such curves.

Fingerlings of a particular species of fish were put into four tanks and kept at temperatures of 25, 27, 29, and 31 degrees Celsius. After fourteen days, and weekly thereafter, one fish was randomly selected from each tank and its length measured. The data from this experiment are given in **Output 5.1**.

Output 5.1 Data Used to Estimate a Growth Curve

TEMP AGE	25	27	29	31
14	620	625	590	590
21	910	820	910	910
28	1315	1215	1305	1205
35	1635	1515	1730	1605
42	2120	2110	2140	1915
49	2300	2320	2725	2035
56	2600	2805	2890	2140
63	2925	2940	3685	2520
70	3110	3255	3920	2710
77	3315	3620	4325	2870
84	3535	4015	4410	3020
91	3710	4235	4485	3025
98	3935	4315	4515	3030
105	4145	4435	4480	3025
112	4465	4495	4520	3040
119	4510	4475	4545	3177
126	4530	4535	4525	3180
133	4545	4520	4560	3180
140	4570	4600	4565	3257
147	4605	4600	4626	3166
154	4600	4600	4566	3214

You can estimate the growth curve for the fish kept at 29 degrees by using a fourth degree polynomial for the independent variable AGE.

The polynomial regression is implemented using PROC REG with LENGTH as the dependent variable and AGE, AGE^2, AGE^3, and AGE^4 as the independent variables. In order to do this, you must first generate the values of the powers of AGE in the DATA step using the following statements:

```
DATA FISH;
   INPUT AGE LENGTH;
   ASQ=AGE*AGE;
   ACUB=AGE*AGE*AGE;
   AQT=ASQ*ASQ;
```

The implementation of the regression requires the following statements:

```
PROC REG;
   MODEL LENGTH=AGE ASQ ACUB AQT / SS1 SEQB;
```

The results of these statements are presented in **Output 5.2**.

Output 5.2 Polynomial Regression Using PROC REG

```
          ❹
SEQUENTIAL PARAMETER ESTIMATES

INTERCEP  3524.62
AGE        1143.96   28.3412
ASQ       -816.642  90.9691 -.372785
ACUB      -1037.46  103.279 -.545668  6.9E-04
AQT        280.188 -2.80282  1.95644 -.021584  6.6E-05

DEP VARIABLE: LENGTH
                                     ANALYSIS OF VARIANCE

                                     SUM OF          MEAN                          ❶
                SOURCE      DF      SQUARES        SQUARE        F VALUE        PROB>F

                MODEL        4  38250790.38    9562697.59        724.314        0.0001
                ERROR       16    211238.57   13202.41083
                C TOTAL     20  38462028.95

                   ROOT MSE       114.9017      R-SQUARE        0.9945
                   DEP MEAN       3524.619      ADJ R-SQ        0.9931
                   C.V.           3.259976

                                     PARAMETER ESTIMATES
                        ❸
                     PARAMETER       STANDARD     T FOR H0:                              ❷
   VARIABLE  DF       ESTIMATE          ERROR   PARAMETER=0     PROB > |T|       TYPE I SS

   INTERCEP   1      280.18822      287.02299         0.976         0.3435       260881728
   AGE        1    -2.80282479    20.35327746        -0.138         0.8922     30305600.11
   ASQ        1     1.95643543     0.45368989         4.312         0.0005      7484972.88
   ACUB       1    -0.02158374    0.003944649        -5.472         0.0001     34496.08880
   AQT        1    0.000066279    0.000011672         5.679         0.0001       425721.31
```

The circled numbers on the output have been added to key the descriptions that follow:

1. You will note that the test for the entire model is statistically significant since the *p* value for the test for MODEL is less than 0.0001. The large value for the coefficient of determination (R-SQUARE) of 0.9945 shows that this model accounts for a large portion of the variation in fish lengths. The residual standard deviation (ROOT MSE), 114.9017, indicates how well the fourth degree polynomial curve fits the data.

   The statistics above do not show if the full fourth degree polynomial is really needed. To answer this question, you use the Type I sums of squares, which are also often called the *sequential* sums of squares.
2. These statistics, labeled TYPE I SS, represent the contribution of each independent variable to the regression sum of squares as that variable is added to the model in the order listed in the MODEL statement (see section **2.4.2**). Dividing these Type I sums of squares by the residual mean square provides *F* statistics, which are then used to test if these additional contributions to the regression sum of squares justify addition of the corresponding terms to the model. In other words, the Type I sums of squares provide the vehicle to determine the degree of polynomial required to describe the relationship adequately.

   In this example, the Type I SS for the linear regression on AGE is 30305600; dividing by the residual mean square of 13202 gives an *F* ratio of 3395.46, which clearly establishes that a linear regression fits better than a regression that includes only an intercept.*

* The Type I SS for INTERCEP is the correction for the mean. This can be used to test the hypothesis that the mean is zero; this test is not usually of interest.

The additional contribution of the quadratic term, designated ASQ, is 7484973, and the F ratio is 566.94; therefore, you can justify adding the quadratic term. The F ratio for adding the cubic term is 2.61, and you cannot justify adding that term. Nevertheless, continue to check for additional terms. The fourth degree term adds 425721, and the F ratio is 32.25; hence, the fourth degree term should be included. You could, of course, continue if you had specified a higher order polynomial in the MODEL statement. However, polynomials beyond the fourth degree are not often used.

3. In this example it is appropriate to recommend the fourth degree polynomial. The estimated equation, obtained from the portion of the output labeled PARAMETER ESTIMATE, is as follows:

$$\text{LENGTH} = 280.19 - 2.8028(\text{AGE}) + 1.9564(\text{AGE})^2 - 0.02158(\text{AGE})^3 + 0.00006628(\text{AGE})^4 \ .$$

The remainder of the statistics for the coefficients reflect the contribution of these terms to the model in the presence of all other coefficients. Since you do not normally consider models that do not contain all terms with powers less than the degree of polynomial, these remaining statistics are of little interest.

4. If the tests based on the Type I sums of squares had indicated that a lower order polynomial would suffice, the coefficients for such lower order polynomial regression models can be found under the heading SEQUENTIAL PARAMETER ESTIMATES produced by PROC REG using the MODEL option SEQB. In this portion of the output, the first line is the zero order polynomial, that is, the mean of the dependent variable (3524.62). The second line contains the coefficients of the first order, or linear, regression:

$$1143.96 + 28.3412(\text{AGE}) \ .$$

The third line contains the coefficients of the quadratic model, and so forth. The last line contains the coefficients of the full or, in this case, the fourth order polynomial and is the same as the list of coefficients under PARAMETER ESTIMATE discussed above in 3.

5.3 POLYNOMIAL PLOT

Using sequential testing to obtain the appropriate degree of polynomial required to fit the growth curve, you could see that while the fourth degree term was required, it was also evident that the major improvement in fit occurred with the addition of the quadratic term. The reduction in residual mean square was much less dramatic for the addition of cubic and fourth degree terms. A plot illustrating how the fit of a polynomial regression improves with the addition of higher order terms may provide information to help you decide if the improvement due to adding additional terms is indeed worthwhile.

In order to construct such a plot, you first need the predicted values associated with the linear, quadratic, cubic, and fourth order regression model estimates. This is accomplished by using the following SAS statements:

```
PROC REG DATA=FR4X1;
   MODEL LENGTH=AGE;
   OUTPUT OUT=L P=PL R=RL;
   MODEL LENGTH=AGE ASQ;
   OUTPUT OUT=Q P=PQ R=RQ;
   MODEL LENGTH=AGE ASQ ACUB;
   OUTPUT OUT=C P=PC R=RC;
   MODEL LENGTH=AGE ASQ ACUB AQT;
   OUTPUT OUT=QT P=PQT R=RQT;
DATA ALL;
   MERGE L Q C QT;
```

The four MODEL statements are for the linear, quadratic, cubic, and fourth order polynomial models, respectively. The four OUTPUT statements create data sets with the predicted and residual values for these models; the predicted values are PL, PQ, PC, and PQT, respectively. The residual values RL, RQ, RC, and RQT will be used later. The MERGE statement places all of these into one data set called ALL.

The desired plot is produced using the following statements:

```
PROC PLOT DATA=ALL;
   PLOT LENGTH*AGE='*' PL*AGE='L' PQ*AGE='Q' PC*AGE='C'
        PQT*AGE='4' / OVERLAY;
```

Unfortunately, the resulting printer plot does not possess sufficient resolution to illustrate the improvement in fit for the cubic and fourth order polynomial terms, and it is not reproduced here. Instead, the equivalent plot produced by PROC GPLOT, which is available with SAS/GRAPH software, is shown in **Output 5.3** using the cubic spline interpolation for the fitted curves.

The instructions needed to produce this plot, not including those that are hardware specific, are as follows:

```
PROC GPLOT;
   BY TEMP;
   PLOT LENGTH*AGE=1 PL*AGE=2 PQ*AGE=3 PC*AGE=4 PQT*AGE=5 / OVERLAY;
   SYMBOL2 V=L I=SPLINE C=BLACK;
   SYMBOL3 V=Q I=SPLINE C=BLACK;
   SYMBOL4 V=C I=SPLINE C=BLACK;
   SYMBOL5 V=4 I=SPLINE C=BLACK;
```

Output 5.3 Using PROC GPLOT to Plot Polynomial Terms

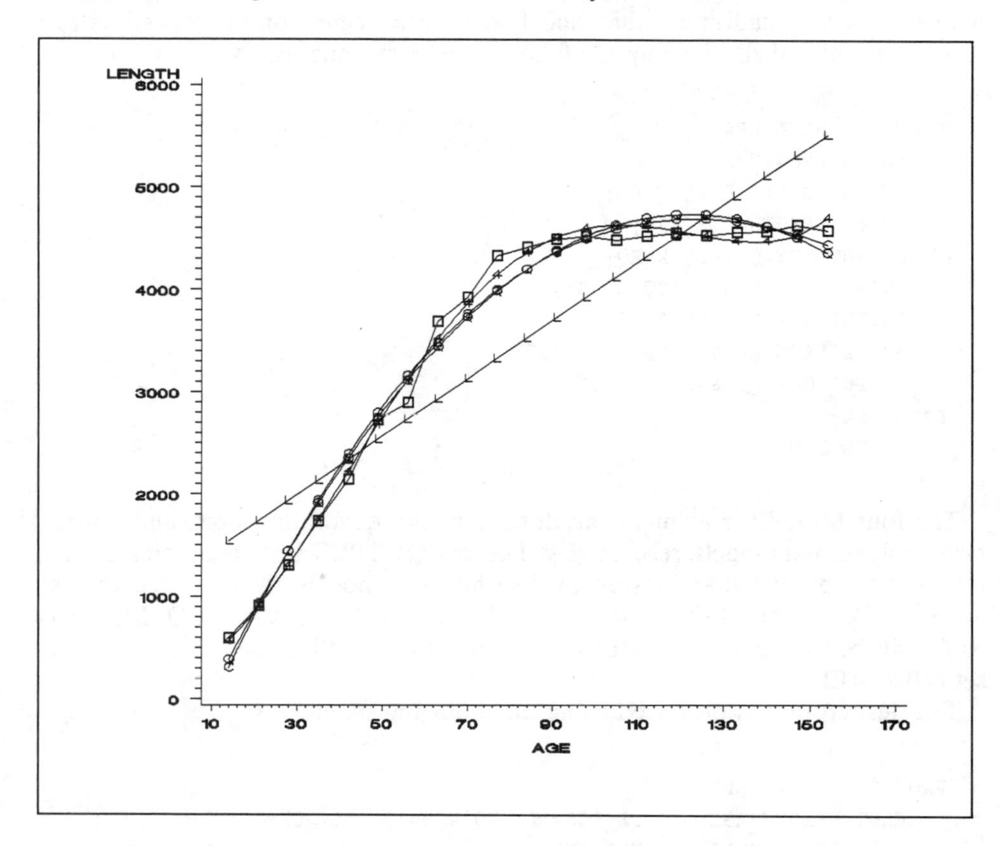

This plot clearly shows the limited improvement due to the cubic and fourth degree terms. In fact, the fourth order polynomial curve shows a peculiar hook at the upper end that is not typical of growth curves. Of course, the quadratic curve shows negative growth in this region, so it is also unsatisfactory.

Another method for checking the appropriateness of a model is to plot the residual values. The plot of residual values from the quadratic polynomial (for which the predicted and residual values are in data set Q) is obtained with the following statements:

```
PROC PLOT DATA=Q;
   PLOT RQ*AGE / VREF=0;
```

The resulting plot, shown in **Output 5.4**, shows a systematic pattern typical of residual plots when the specified degree of polynomial is inadequate. In this case the W-shaped pattern is due to the fourth degree term that was statistically significant.

Output 5.4 A Plot of Residual Values Using PROC PLOT

```
            PLOT OF RQ*AGE    SYMBOL USED IS *

       I
   400 +
       I
       I                      *
       I
   300 +
       I
       I                  *
       I                        *                   *
   200 +    *               *
R      I
E      I
S      I                                          *
I  100 +                          *
D      I
U      I
A      I
L    0 +----------------------------*---------------------
S      I      *       *
       I                                        *
       I
  -100 +
       I        *                             *
       I                              *
       I          *                     * *
  -200 +            *                       *
       I                *
       I
       I
  -300 +
       -----+-----+-----+-----+-----+-----+-----+---------
           14    35    56    77    98   119   140

                             AGE
```

5.4 POLYNOMIAL MODELS WITH SEVERAL VARIABLES

Polynomial models for several variables contain terms involving powers of the various variables as well as cross products among these variables. The cross product terms are used to detect interactions among the effects of the variables. For example, consider the following two-variable model:

$$y = \beta_0 + \beta_1 x_1 + \beta_2 x_2 + \beta_{12} x_1 x_2 \quad .$$

The coefficient β_{12} indicates how the linear effect of x_1 is affected by x_2 or vice versa. This is seen by the following rearrangement of terms:

$$y = \beta_0 + (\beta_1 + \beta_{12} x_2) x_1 + \beta_2 x_2 \quad .$$

The coefficient β_{12} specifies how the linear effect of x_1 changes with x_2. This change is linear in x_2; hence, this effect is called the linear-by-linear interaction. The interaction effect is symmetric. It also shows how the linear coefficient in x_2 is affected by x_1. Cross products involving higher order terms have equivalent connotation, although their interpretation may become more difficult.

Regressions using such models can be implemented by generating the desired powers and cross products variables in the DATA step and implementing a multiple regression analysis using PROC REG.

One consequence of the greater complexity of models with polynomials in several variables is that the sequential (Type I) sums of squares are no longer useful in selecting the appropriate degree of model. The partial (Type II) sums of squares are also of little use since, as in the case for one-variable models, it is not customary to omit lower order terms. A similar restriction also applies to cross product terms. For example, if a cross product of two linear terms has been included, you should also include the individual linear terms. In such models, you are primarily interested in performing tests to answer the following questions:

- Does the entire model help to explain the behavior of the response variable?
- Do you need all the variables or factors?
- Do you need quadratic or higher order terms?
- Do you need cross product terms?
- Is the model adequate?

Obviously, the statistics supplied by a single run of a regression analysis, such as the output of PROC REG, cannot answer all your questions.

Since polynomial models with several variables can easily become cumbersome and consequently difficult to interpret, it is common practice to restrict the degree of polynomial terms used for such models. The most frequently used model of this type is called the *quadratic response surface model* in which the maximum total exponent of any term is two. In other words, this model includes all linear and quadratic terms in the individual variables and all pairwise cross products of linear terms. PROC RSREG (for Response Surface REGression) is a procedure used for building a quadratic response surface model. The implementation of this procedure is illustrated by estimating the response surface regression relating the LENGTH of fish to AGE and TEMP using the data in **Output 5.1**.

The procedure is implemented with the following statements:

```
PROC RSREG DATA=FISH;
   MODEL LENGTH=AGE TEMP;
```

Note that the MODEL statement requires only the listing of the dependent and independent variables or factors; PROC RSREG creates the necessary squares and product variables. For this example, the regression model estimated by the procedure is as follows:

$$\begin{aligned}\text{LENGTH} = \beta_0 &+ \beta_1(\text{AGE}) + \beta_2(\text{AGE})^2 + \beta_3(\text{TEMP}) \\ &+ \beta_4(\text{TEMP})^2 + \beta_5(\text{AGE})(\text{TEMP}) \quad .\end{aligned}$$

Options that create an output data set containing predicted, residual, and other statistics associated with individual observations are available in PROC RSREG. These options are discussed in section **5.6**. One option that is not illustrated is the COVARIATES option, which allows the introduction of variables that are not part of the response surface model. This option is useful for adjusting estimates for experimental conditions.

The results of the implementation of PROC RSREG on the fish data are given in **Output 5.5**.

Output 5.5 A Quadratic Response Surface Model Produced with PROC RSREG

```
RESPONSE SURFACE FOR VARIABLE LENGTH

                          RESPONSE MEAN         3153.345
                          ROOT MSE              262.0385
                          R-SQUARE             0.9610963
                          COEF OF VARIATION   0.08309857

                          REGRESSION       DF    TYPE I SS     R-SQUARE   F-RATIO    PROB

                      ❷   LINEAR            2    107867300       0.7835    785.47  0.0001
                          QUADRATIC         2  21762116.20       0.1581    158.47  0.0001
                      ❸   CROSSPRODUCT      1   2682926.41       0.0195     39.07  0.0001
                          TOTAL REGRESS     5    132312342       0.9611 ❶  385.39  0.0001

                          RESIDUAL         DF           SS  MEAN SQUARE

                          TOTAL ERROR      78   5355804.57  68664.16119

                          PARAMETER        DF     ESTIMATE      STD DEV   T-RATIO    PROB

                          INTERCEPT         1 -56024.90172   5625.61404     -9.96  0.0001
                          AGE               1    123.82245   8.98830676     13.78  0.0001
                          TEMP              1   3934.86290    401.27536      9.81  0.0001
                      ❹   AGE*AGE           1  -0.26670820   0.01785248    -14.94  0.0001
                          TEMP*AGE          1  -1.88558442   0.30165231     -6.25  0.0001
                          TEMP*TEMP         1 -69.20535714   7.14768533     -9.68  0.0001

                          FACTOR           DF           SS  MEAN SQUARE   F-RATIO    PROB

                          AGE               3    121756815     40585605    591.07  0.0001
                          TEMP              3     13238453      4412818     64.27  0.0001

SOLUTION FOR OPTIMUM RESPONSE

         FACTOR CRITICAL VALUE

   ❺     AGE         138.29700
         TEMP      26.54485424

PREDICTED VALUE AT OPTIMUM       4762.416

EIGENVALUES   EIGENVECTORS
                      AGE          TEMP
  -0.253817     0.9999065     -0.013672
   -69.2182    0.01367198     0.9999065

SOLUTION WAS A MAXIMUM
```

The five questions posed in the introduction to this section can be answered by studying **Output 5.5**. The circled numbers on the output have been added to key the descriptions that follow:

1. The significance of the overall model is derived from the statistics on this row. The computed F ratio of 385.39 readily leads to the rejection of the hypothesis that the model is ineffective in explaining the variation seen in length.
2. The need for linear and quadratic terms is provided by the statistics in these two rows. The line labeled LINEAR tests the effectiveness of the model including only the strictly linear terms. The line labeled QUADRATIC tests the additional contribution of the quadratic terms, $(AGE)^2$ and $(TEMP)^2$. Obviously, some linear and quadratic terms are needed, although these statistics do not specify which of these you need.
3. The need for the CROSSPRODUCT terms (in this case there is only one) is provided in this line. Obviously, this term should also be included in the model.

4. The contribution of the variables, or factors, is provided from these lines. In this section, the line labeled AGE tests the hypothesis that the factor AGE can be omitted from the model. The resulting *F* ratio tests the increase in the residual mean square if all terms involving AGE, namely AGE, $(\text{AGE})^2$, and (AGE)(TEMP), are deleted from the model. Obviously, the factor AGE should not be omitted from the model. Similarly, the line labeled TEMP determines if the factor TEMP can be omitted. Although this factor contributes less than AGE, it also appears to be needed to describe the growth of the fish.
5. Since response surface analysis is sometimes performed to obtain information on optimum estimated response, PROC RSREG also supplies information to assist in determining if the estimated response surface exhibits such an optimum. This information is presented in the section labeled SOLUTION FOR OPTIMUM RESPONSE. First, the partial derivatives of the estimated response surface equation are calculated, and a stationary point is found. The values of the factors for the stationary point are given as FACTOR CRITICAL VALUES, and the estimated response at this point is denoted PREDICTED VALUE AT OPTIMUM. For this example, the stationary point is 138.3 for AGE and 25.545 for TEMP, where the estimated response is 4762.4. The procedure does not check to see if these values are in the range of data. Because the stationary point may be a maximum, minimum, or saddle point, a canonical analysis is performed to ascertain which it is. In this case the point is a maximum; the eigenvalues and eigenvectors required for the analysis may yield additional information on the shape of the response surface (Myers 1976).

Additional statistics in **Output 5.5** include some overall descriptive measures, the residual sum of squares and the residual mean square. The various statistics for the individual coefficients are given in the section under the headings that start with PARAMETER. Here you can see, for example, that all linear and quadratic coefficients are statistically significant. As before, the statistics for the lower order terms are not particularly useful.

Although the model fits the data reasonably well, you do not know with certainty that the model is adequate. In order to do this, an independent estimate of the true error variance is needed. An estimate of this variance is not available for these data, so the test cannot be made for the output in **Output 5.5**. Such a test is discussed in section **5.6**.

5.5 RESPONSE SURFACE PLOTS

As in the case of a one-variable polynomial regression, graphic representations are very useful to depict the nature of the estimated curve. For multidimensional polynomial regressions, such plots are called *response surface plots.* A popular plot of this type is a contour plot, where contours of equal response are plotted for a grid of values of the independent variables. Such a plot can be implemented with PROC PLOT. Also available are three-dimensional representations that can be performed using PROC G3D in SAS/GRAPH software.

As an example of a response surface plot, a plot is made of the estimated quadratic response surface for the fish lengths produced in the previous section. In addition, a residual plot that may be useful for investigating specification error or detecting possible outliers is provided.

All SAS System plotting procedures construct plots that demonstrate relationships among values of variables in a data set. Therefore, in order to produce a response surface plot, it is first necessary to produce a data set of values representing the estimated response for a grid of values of the independent variables to be used in the plot. This is not difficult since all SAS System regression procedures allow the calculation of predicted values for data points that are not in the data set used to estimate the regression model.

First, generate the grid of values of the independent variables. The number of grid points required depends on the size of the page used for the computer output. In this example, the page size allows 100 columns and 80 rows. Allowing for legends and titles, a plot of about 60 rows (representing TEMP) and 60 columns (representing AGE) is used for this plot. Plots involving more than two independent variables will be discussed later.

The grid of values is generated in the data set F1 with the following statements:

```
DATA F1;
   DO TEMP=25 TO 31 BY .1;
      DO AGE=10 TO 160 BY 2.5;
         ID=1;
         OUTPUT;
      END;
   END;
```

The data set F1 has 61 x 61=3,721 observations, containing the variables TEMP and AGE and a variable ID whose value is unity; the need for this variable is discussed later in this chapter. Next, combine data set F1 with the original data set:

```
DATA F2;
   SET FISH F1;
```

Note that data set FISH does not have the variable ID and the data set F1 does not have the variable LENGTH. These variables are denoted as missing in data set F2. PROC RSREG is used with data set F2, and statements are used to produce the predicted values for the grid of independent variable values.*

* The data set F2 may be created in a different and possibly more efficient manner by implementing the EOF option in the DATA step. For an example, see the *SAS User's Guide: Statistics, Version 5 Edition.*

The format and options for producing estimated values (and other statistics) are different for PROC RSREG than for PROC REG or PROC GLM. The following statements are required:

```
PROC RSREG DATA=F2 OUT=P1;
   BY ID;
   MODEL LENGTH=AGE TEMP / PREDICT RESIDUAL BYOUT;
```

The PROC statement must include the option OUT=P1, which specifies that the output data set containing the predicted values is data set P1. The BY ID statement, together with the BYOUT option, specifies that PROC RSREG uses only the observations of the first BY group for estimation of parameters. Remember that ID has missing values for the actual data set and value of unity for the grid set. The PROC uses the observed data for estimating because ID=. sorts before ID=1. If the BY statement read

```
BY ID NOTSORTED;
```

it would use the observed data because it was placed first when F2 was created.

The MODEL statement options PREDICT and RESIDUAL request that the output data set contains the predicted and residual values; other statistics, such as the actual values, confidence intervals, and the Cook's *D* statistic, may be obtained by adding other keywords (see the *SAS User's Guide: Statistics, Version 5 Edition*).

Output data sets produced by PROC RSREG have a different structure from those produced by PROC REG. The first twenty observations from the data set produced with the SAS statements given above are reproduced in **Output 5.6**.

Output 5.6 Output Data Set Produced by PROC RSREG

OBS	ID	AGE	TEMP	_TYPE_	LENGTH
1	.	14	25	PREDICT	114.61
2	.	14	25	RESIDUAL	505.39
3	.	21	25	PREDICT	586.04
4	.	21	25	RESIDUAL	323.96
5	.	28	25	PREDICT	1031.34
6	.	28	25	RESIDUAL	283.66
7	.	35	25	PREDICT	1450.50
8	.	35	25	RESIDUAL	184.50
9	.	42	25	PREDICT	1843.53
10	.	42	25	RESIDUAL	276.47
11	.	49	25	PREDICT	2210.42
12	.	49	25	RESIDUAL	89.58
13	.	56	25	PREDICT	2551.16
14	.	56	25	RESIDUAL	48.84
15	.	63	25	PREDICT	2865.78
16	.	63	25	RESIDUAL	59.22
17	.	70	25	PREDICT	3154.25
18	.	70	25	RESIDUAL	-44.25
19	.	77	25	PREDICT	3416.59
20	.	77	25	RESIDUAL	-101.59

The data set, named P1, contains the following variables:

- the independent variables or factors (AGE and TEMP)
- variables specified by either ID or BY statements (ID)
- a variable called _TYPE_, whose use is described below
- one or more variables labeled by the names of the dependent variables.

There is one observation in **Output 5.6** for each statistic requested in the MODEL option for each observation in the input data set. The automatic variable _TYPE_ identifies which statistic each observation represents.

In this example, PREDICT and RESIDUAL are the statistics specified in the MODEL statement. There are two observations for each observation in the input data set. One observation, having _TYPE_=PREDICT, has the predicted value of the dependent variable identified by the name of the dependent variable. The second line, identified by _TYPE_=RESIDUAL, contains the residual value. If there are several dependent variables in the MODEL statement, the corresponding output statistics will have the names of the respective variables.

The format for the data set P1 is not directly suitable for the desired plots. This is because the data set for the residual plot must contain the predicted and residual values for the originally observed data, whereas the data set for the contour plot requires the values of the independent variables and predicted values of the *grid data set*. These data sets are created with the following statements:

```
DATA PLOT PRED (RENAME=(LENGTH=PRED))
          RESID (RENAME=(LENGTH=RESID));
   SET P1;
   IF ID=1 AND _TYPE_='PREDICT' THEN OUTPUT PLOT;
   IF ID=. THEN DO;
      IF _TYPE_='PREDICT' THEN OUTPUT PRED;
      IF _TYPE_='RESIDUAL' THEN OUTPUT RESID;
   END;
DATA RESID2;
   MERGE PRED RESID;
```

The data set for the response surface plot, which is named PLOT, corresponds to the original set F1 in which the ID variable was set to unity.

The residual plot requires the original data set in which the variable ID is missing. The first DATA step creates two data sets with the predicted and residual values labeled PRED and RESID, respectively. These data sets are then merged to provide the data sct RESID2, which is needed for the residual plot. The residual plot is obtained with the following statements:

```
PROC PLOT DATA=RESID2;
   PLOT RESID*PRED / VREF=0;
```

The plot is shown in **Output 5.7**. Although there seem to be patterns in these residuals, the patterns are not consistent enough to identify specification errors.

Output 5.7 A Plot of Residual Values

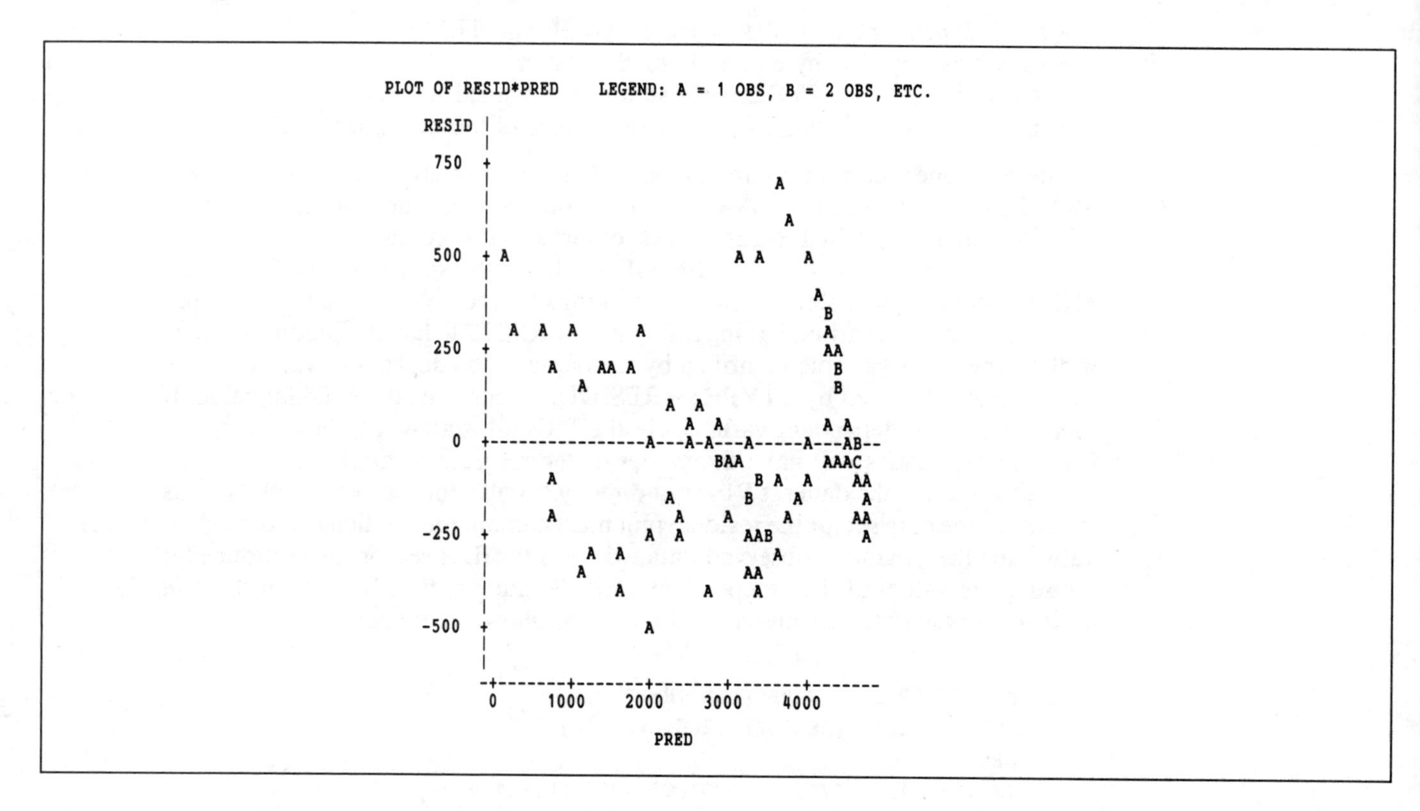

You can produce the contour plot with the following statements:

```
PROC PLOT DATA=PLOT;
   PLOT TEMP*AGE=LENGTH / CONTOUR=8;
```

The PLOT statement specifies that TEMP is the rows, AGE the columns, and that the variable LENGTH, which contains the value predicted by the response surface regression, defines the values generating the contours. The keyword CONTOUR=8 specifies that you want a contour plot with eight levels of contours. The resulting plot is shown in **Output 5.8**. The plot shows the growth of fish with age. The growth becomes negative with increasing age. The plot also shows that fish grow faster at temperatures of 26 to 28 degrees Celsius.

Output 5.8 Using the Contour Option with PROC PLOT

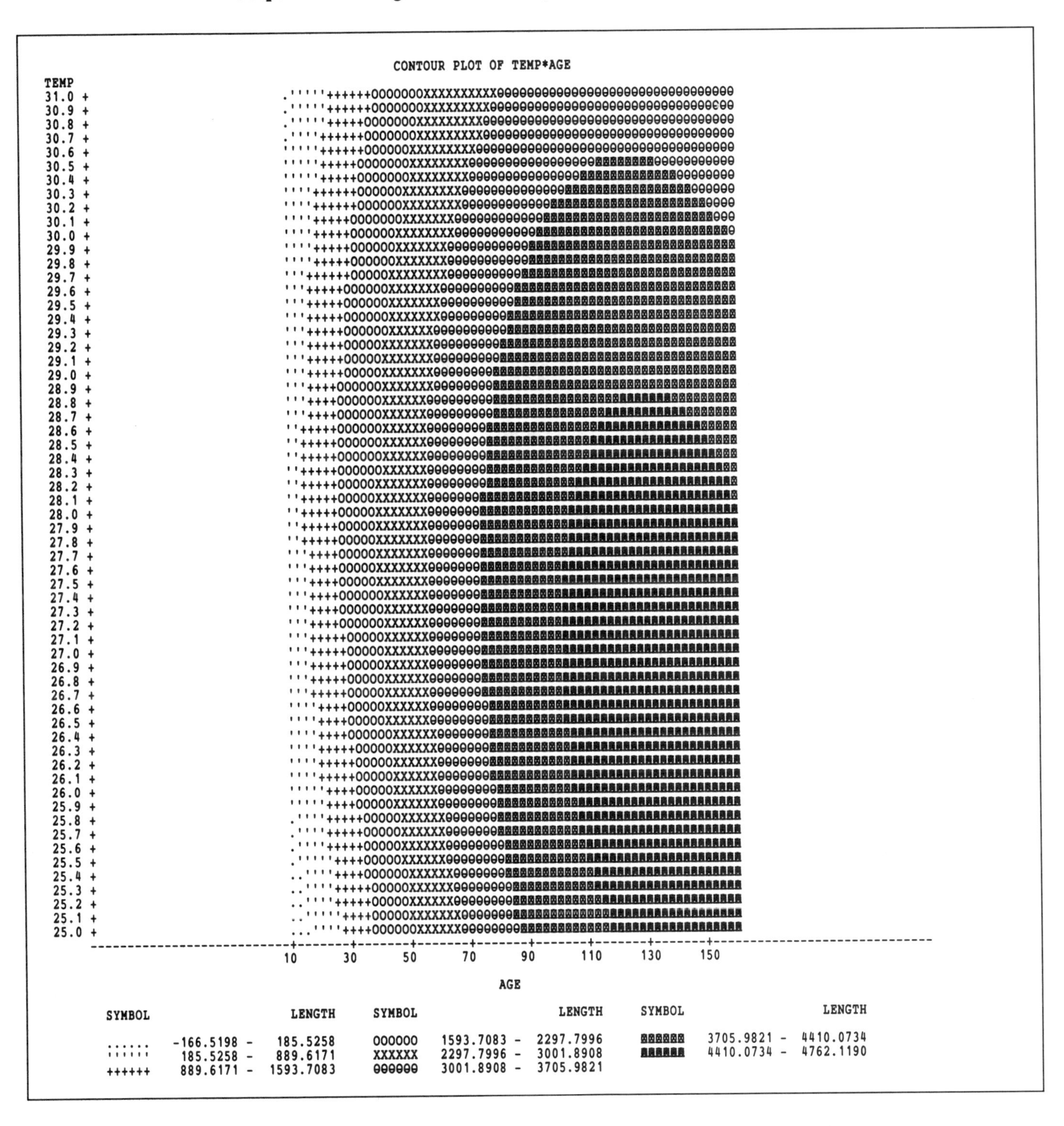

5.6 A THREE-FACTOR RESPONSE SURFACE EXPERIMENT

The data for this example come from an experiment concerning a device for automatically shelling peanuts (Dickens and Mason 1962).

> Peanuts flow through stationary sheller bars and rest on the grid which has perforations just large enough to pass shelled kernels. The grid is reciprocated and the resulting forces on the peanuts between the moving grid and the stationary bars break open the hulls. . . . The problem became one of determining the combination of bar grid spacing, length of stroke and frequency of stroke which would produce the most satisfactory performance. The performance criteria are . . . kernel damage, shelling time and unshelled peanuts.

The paper cited above describes three separate experiments; for this example, only the data from the second experiment are used. The experimental design is a three-factor composite design consisting of fifteen points, with five additional observations at the center point (Myers 1976). The data consist of responses resulting from the shelling of 1000 grams of peanuts. The factors of the experiment are as follows:

LENGTH is the length of stroke (inches).
FREQ is the frequency of stroke (strokes/minute).
SPACE is the bar grid spacing (inches).

The response variables are as follows:

TIME is time to shell 1000 grams of peanuts (minutes).
UNSHL is grams of unshelled peanuts.
DAMG is percentage of damaged peanuts.

The data are presented in **Output 5.9**.

Output 5.9 Data for Response Surface Analysis

OBS	LENGTH	FREQ	SPACE	TIME	UNSHL	DAMG
1	1.00	175	0.86	16.00	284	3.55
2	1.25	130	0.63	9.25	149	8.23
3	1.25	130	1.09	18.00	240	3.15
4	1.25	220	0.63	4.75	155	5.26
5	1.25	220	1.09	15.50	197	4.23
6	1.75	100	0.86	13.00	154	3.54
7	1.75	175	0.48	3.50	100	8.16
8	1.75	175	0.86	7.00	176	3.27
9	1.75	175	0.86	6.25	177	4.38
10	1.75	175	0.86	6.50	212	3.26
11	1.75	175	0.86	6.50	200	3.57
12	1.75	175	0.86	6.50	160	4.65
13	1.75	175	0.86	6.50	176	4.02
14	1.75	175	1.23	12.00	195	3.80
15	1.75	250	0.86	5.00	126	4.05

(continued on next page)

(continued from previous page)

```
16    2.25    130    0.63    4.00     84    9.02
17    2.25    130    1.09    7.00    145    3.00
18    2.25    220    0.63    2.25     97    7.41
19    2.25    220    1.09    5.75    168    3.78
20    2.50    175    0.86    3.50    168    3.72
```

Since this experiment has six replications at the center point (LENGTH=1.75, FREQ=175, and SPACE=0.86; observations 8-13), you can obtain an estimate of pure error and consequently perform a test for lack of fit. In this example, only the variable UNSHL is used. You may want to perform analyses for the other responses to compare the results. The following SAS statements are used:

```
PROC SORT;
   BY LENGTH FREQ SPACE;
PROC RSREG;
   MODEL UNSHL=LENGTH FREQ SPACE / LACKFIT;
```

The use of PROC SORT ensures that all observations from the same set of treatment or factor combinations are together. When you sort the data in this manner, the LACKFIT option in PROC RSREG computes a within or pure error sum of squares from all observations occurring within identical factor combinations. This sum of squares is subtracted from the residual sum of squares in the model to obtain the lack-of-fit sum of squares. This quantity indicates the additional variation that can be explained by adding to the model all additional parameters allowed by the structure of the treatment design. Thus, the ratio of the resulting lack-of-fit and pure error mean squares provides a test for the possible existence of such additional model terms. In other words, it is a test for the adequacy of the model. The results of the RSREG procedure are given in **Output 5.10**.

Output 5.10 PROC RSREG Used with the LACKFIT Option

```
RESPONSE SURFACE FOR VARIABLE UNSHL

         RESPONSE MEAN             168.15
         ROOT MSE                19.22284
         R-SQUARE                0.914429
         COEF OF VARIATION      0.1143196

         REGRESSION        DF     TYPE I SS    R-SQUARE   F-RATIO    PROB

         LINEAR             3   27642.45930     0.6401     24.94   0.0001
         QUADRATIC          3   10988.54040     0.2545      9.91   0.0024
         CROSSPRODUCT       3     856.37500     0.0198      0.77   0.5354
         TOTAL REGRESS      9   39487.37469     0.9144     11.87   0.0003

         RESIDUAL          DF            SS MEAN SQUARE   F-RATIO    PROB

      ❶  LACK OF FIT        5    1903.67531   380.73506     1.063   0.4742
         PURE ERROR         5    1791.50000   358.30000
         TOTAL ERROR       10    3695.17531   369.51753
```

(continued on next page)

(continued from previous page)

```
                           PARAMETER          DF      ESTIMATE        STD DEV    T-RATIO     PROB

                           INTERCEPT           1    -130.38663      210.10255      -0.62   0.5488
                           LENGTH              1    -319.78441      111.96408      -2.86   0.0171
                           FREQ                1    3.00670893     1.18000031       2.55   0.0290
                           SPACE               1     789.77558      232.08371       3.40   0.0067
                           LENGTH*LENGTH       1   52.11056417    23.97343299       2.17   0.0548
                           FREQ*LENGTH         1    0.40555556     0.30205777       1.34   0.2091
                        2  FREQ*FREQ           1  -0.009684288    0.002524355      -3.84   0.0033
                           SPACE*LENGTH        1   -1.08695652    59.09825937      -0.02   0.9857
                           SPACE*FREQ          1   -0.47101449     0.65664733      -0.72   0.4896
                           SPACE*SPACE         1    -331.28720      100.06774      -3.31   0.0079

                           FACTOR             DF            SS    MEAN SQUARE    F-RATIO     PROB

                           LENGTH              4      16591.46       4147.864      11.23   0.0010
                        3  FREQ                4      6462.244       1615.561       4.37   0.0266
                           SPACE               4      17546.41       4386.602      11.87   0.0008
```

```
     4
SOLUTION FOR OPTIMUM RESPONSE

         FACTOR CRITICAL VALUE

         LENGTH      2.38162808
         FREQ         179.31274
         SPACE       1.06060235

PREDICTED VALUE AT OPTIMUM             177.1991

EIGENVALUES    EIGENVECTORS
                     LENGTH          FREQ         SPACE
   52.11213       0.9999914   0.003896841   -0.00141991
 -0.0103088     -0.00389784     0.9999922  -0.000704507
   -331.288      0.00141715  0.0007100359     0.9999987

SOLUTION WAS A SADDLE POINT
```

Circled numbers have been added to key the descriptions that follow:

1. First examine the lack-of-fit portion of the results. The residual sum of squares from the nine-term model is 3695.1753, with ten degrees of freedom. The pure error is the sum of squares among the six replicated values for LENGTH=1.75, FREQ=175, and TIME=0.86; the value is 1791.5, with five degrees of freedom. The difference, 1903.6753 with five degrees of freedom, is the additional sum of squares that could be obtained by adding five terms to the model. The *F* statistic derived from the ratio of the lack-of-fit to pure error mean square has a *p* value of 0.4742. You can see that additional terms are not needed.

The remainder of the output is similar to that of the previous example. The results can be summarized as follows:

2. The cross product terms are not significant, so you may conclude that there are no interactions. In other words, the responses to any one factor have similar shapes across levels of the other factors.
3. The factor FREQ has the smallest effect. In fact, it is not significant at the 0.01 level.
4. The response surface has a saddle point; in other words, it has no point at which the response is either maximum or minimum.

Since you are looking for a minimum amount of unshelled peanuts, the shape of the response curve is not too useful. Of course, it is possible that the saddle point is not well-defined and that a broad range of points may provide a guide for finding optimum operating conditions. A plot of the surface may assist in the search for such conditions.

Of course, it is not possible to produce a three-factor response surface plot, but you can produce a plot that represents the response curve for two factors for several levels of the third factor. Because there are three different factor combinations for such plots, you must choose which combination is most useful. Often this decision is not easy to make, and you may have to experiment with several combinations. In this example, the factor FREQ has the smallest effect, so it appears logical to plot the response to LENGTH and SPACE for selected values of FREQ.

With the statements below, a response surface plot will be produced for LENGTH and SPACE for values of FREQ of 175, 200, and 225 strokes per minute. Provisions are also made for a residual plot. As before, you must first generate the data representing the grid of points needed for the response surface plot. The following statements are required:

```
DATA P1;
   ID=1;
   DO FREQ=175 TO 225 BY 25;
      DO LENGTH=1 TO 2.5 BY .03;
         DO SPACE=.6 TO 1.2 BY .01;
            OUTPUT;
         END;
      END;
   END;
```

The data set P1 consists of 51 by 51 grids of values of LENGTH and SPACE for three values of FREQ. The data set P1 is concatenated with the original data set:

```
DATA P2;
   SET PEANUTS P1;
```

The RSREG procedure is implemented:

```
PROC RSREG DATA=P2 OUT=P3;
   MODEL UNSHL=LENGTH FREQ SPACE / PREDICT RESIDUAL BYOUT;
   BY ID;
```

Next, necessary data sets for the plots are created. The procedure is the same as presented in the previous section.

```
DATA PLOT (RENAME=(UNSHL=PREDICT))
     P4 (RENAME=(UNSHL=PREDICT))
     P5 (RENAME=(UNSHL=RESIDUAL));
     SET P3;
   IF ID=1 AND _TYPE_='PREDICT' THEN OUTPUT PLOT;
   IF ID=. AND _TYPE_='PREDICT' THEN OUTPUT P4;
   IF ID=. AND _TYPE_='RESIDUAL' THEN OUTPUT P5;
DATA P6;
   MERGE P4 P5;
```

Data set PLOT contains the predicted values for the grid needed for the response surface plots. Data sets P4 and P5 contain the predicted and residual values for the actual data points since the variable ID was undefined in that set of data. Then sets P4 and P5 are merged to create the data set needed for the residual plot.

The residual plot is constructed with the following statements:

```
PROC PLOT DATA=P6;
   PLOT RESIDUAL*PREDICT / HPOS=35 VPOS=25 VREF=0;
```

The plot is shown in **Output 5.11**.

Output 5.11 A Residual Plot

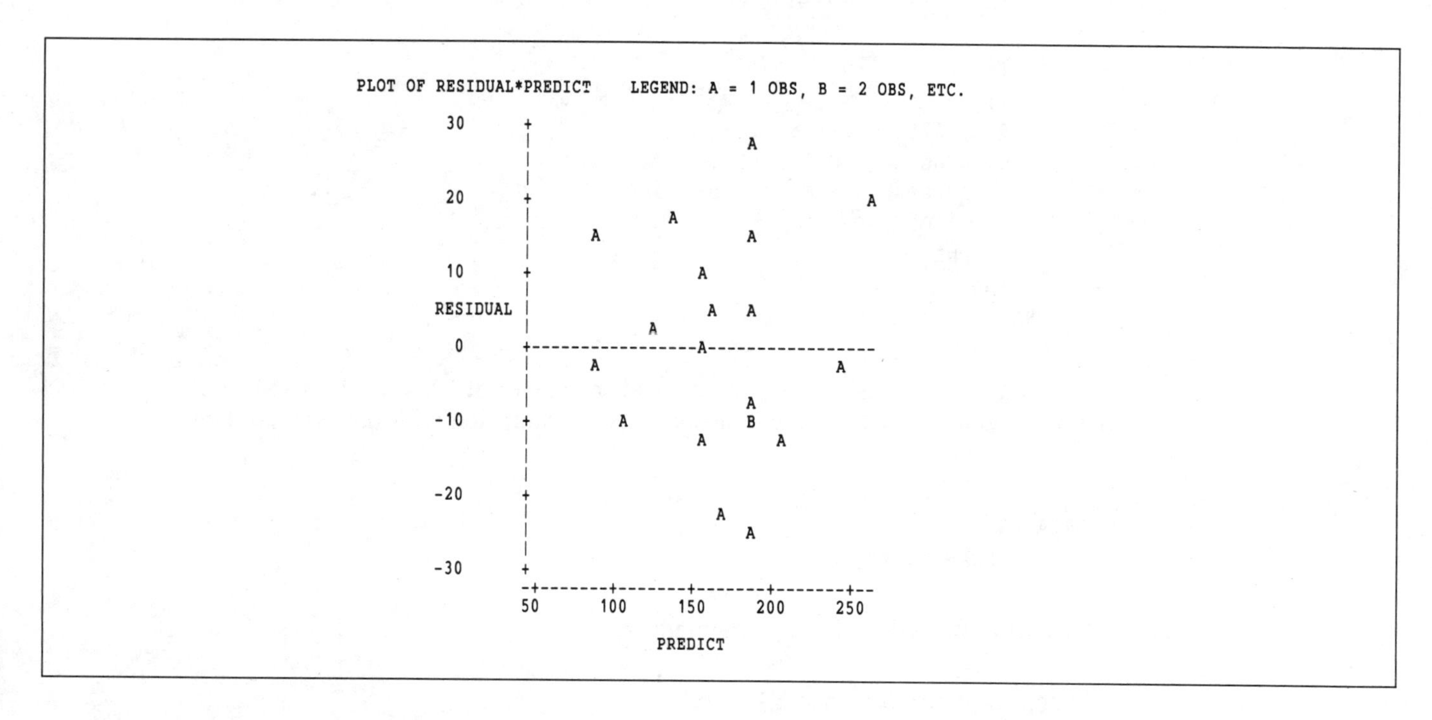

No obvious outliers or specification errors appear in **Output 5.11**, thus confirming the results of the lack-of-fit analysis.

The response surface plots are implemented as before:

```
PROC PLOT DATA=PLOT;
   PLOT LENGTH*SPACE=PREDICT / CONTOUR=6;
   BY FREQ;
```

These plots are shown in **Output 5.12**. You can see the small effect of the FREQ factor and lack of interaction by regarding the similarities of the three plots. The nature of the saddle point becomes clear. Although there is a definite tendency for smaller amounts of unshelled peanuts at low spacing and high length of stroke, the same tendency (although not as strong) is noted for wider spacing. It would seem that the true minimum is outside the range of this experiment; it is probably found at narrower spacing, although the higher spacings should not be ignored. Unfortunately, the region of minimum unshelled peanuts also has the highest damage rates (results not shown here), so some compromise must obviously be sought.

Output 5.12 Three Response Surface Plots

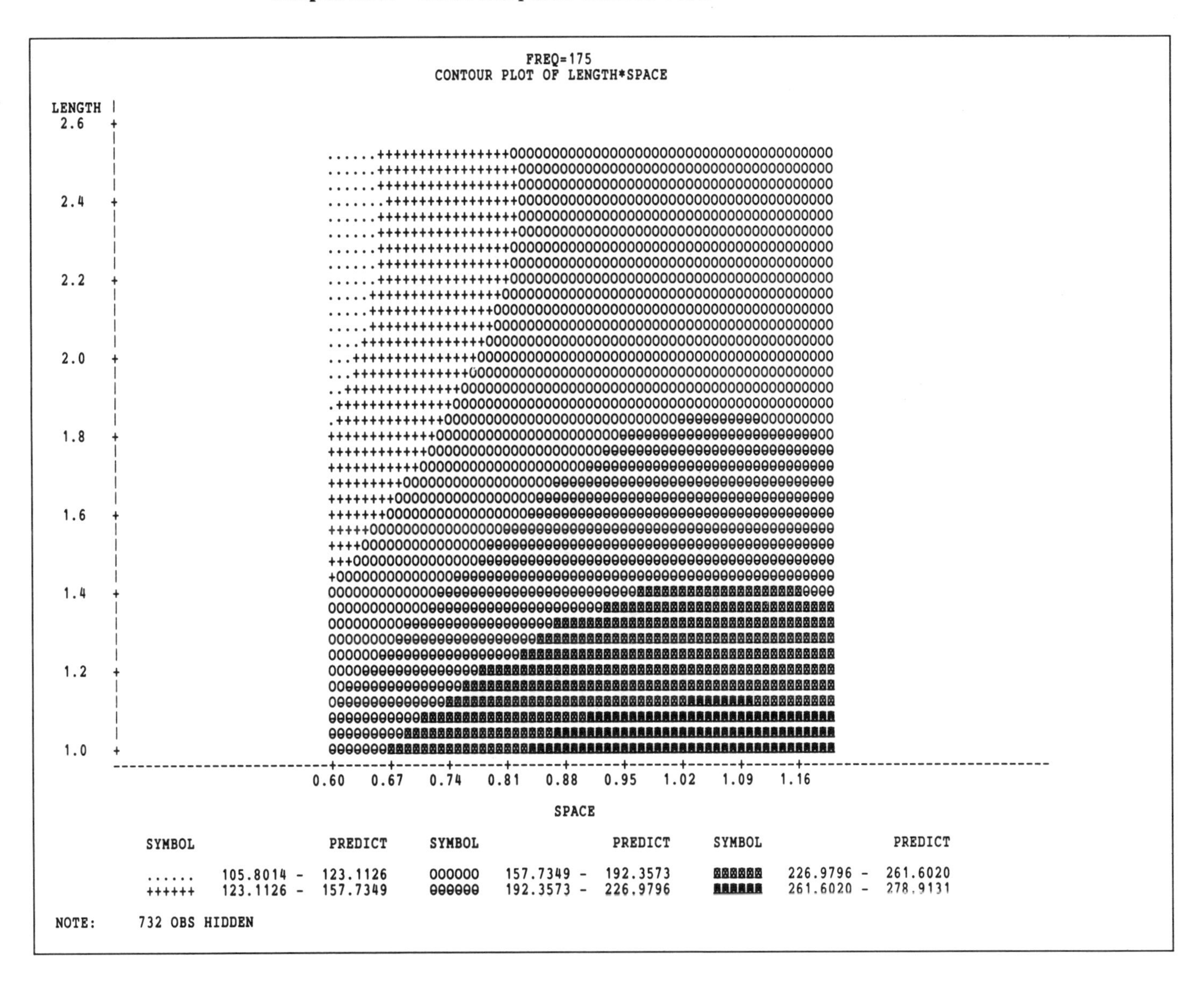

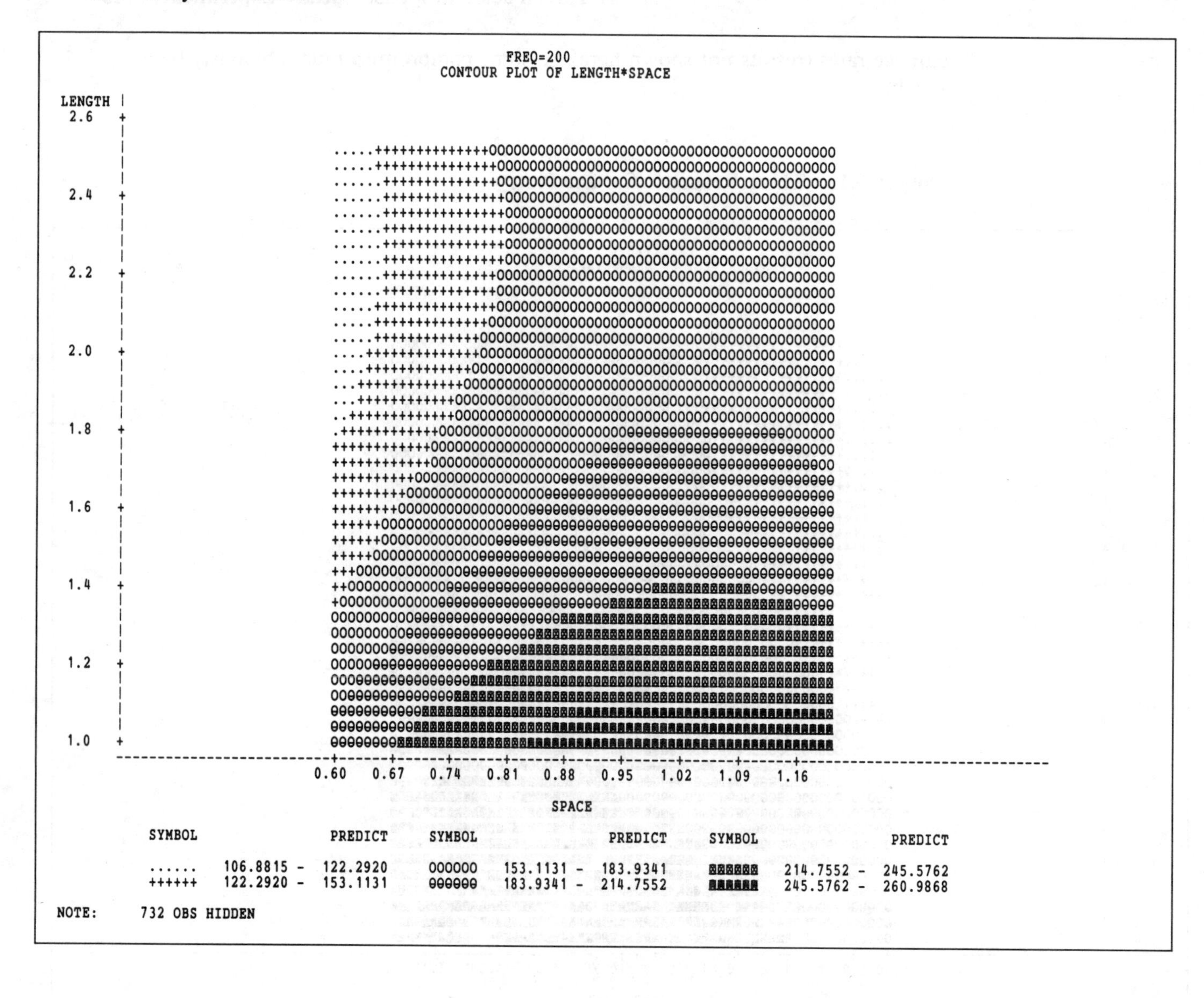
FREQ=200
CONTOUR PLOT OF LENGTH*SPACE
LENGTH
2.6
2.4
2.2
2.0
1.8
1.6
1.4
1.2
1.0
0.60 0.67 0.74 0.81 0.88 0.95 1.02 1.09 1.16
SPACE
SYMBOL PREDICT SYMBOL PREDICT SYMBOL PREDICT
...... 106.8815 - 122.2920
++++++ 122.2920 - 153.1131
000000 153.1131 - 183.9341
183.9341 - 214.7552
214.7552 - 245.5762
245.5762 - 260.9868
NOTE: 732 OBS HIDDEN

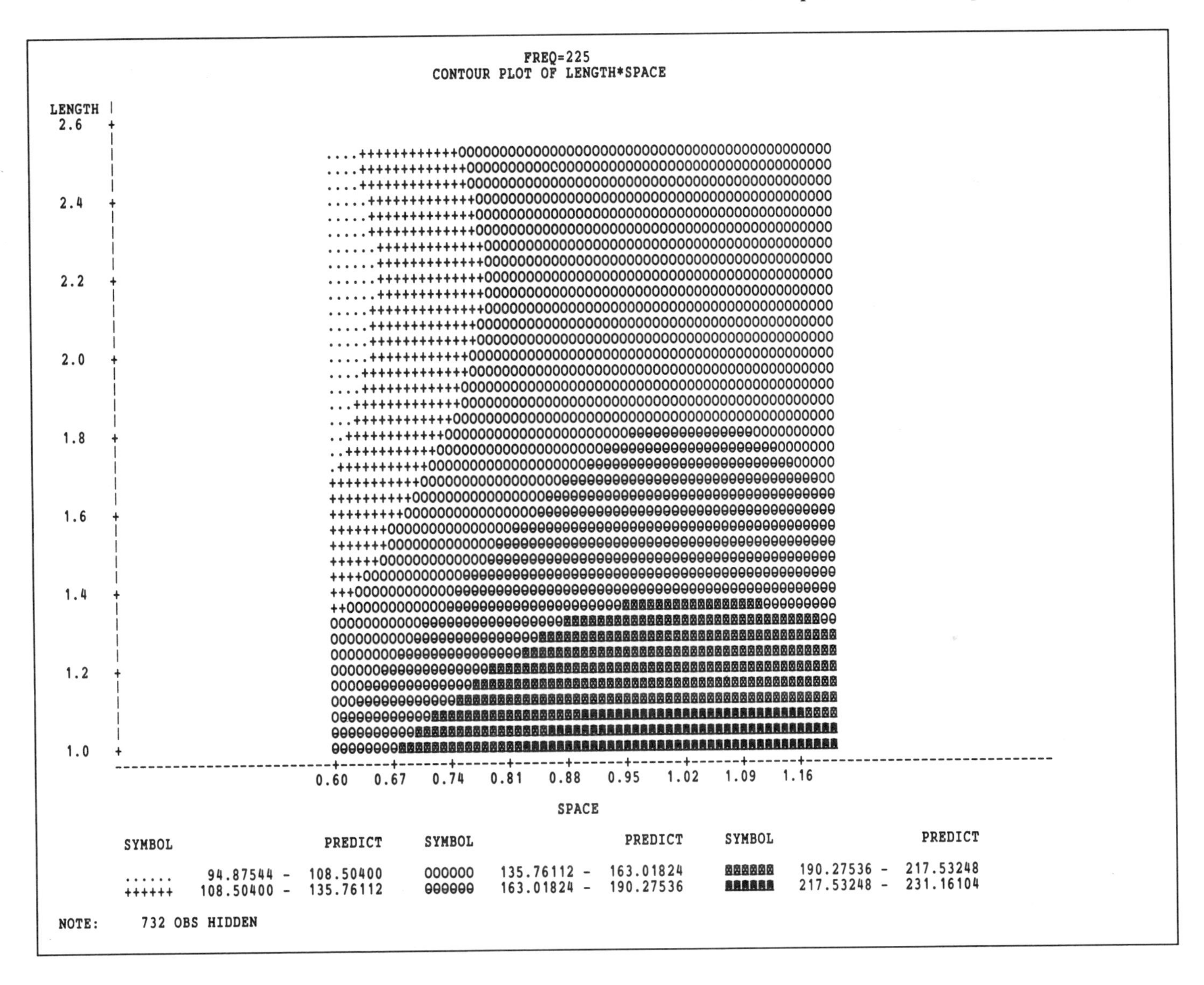
FREQ=225
CONTOUR PLOT OF LENGTH*SPACE
LENGTH
2.6
2.4
2.2
2.0
1.8
1.6
1.4
1.2
1.0
0.60 0.67 0.74 0.81 0.88 0.95 1.02 1.09 1.16
SPACE
SYMBOL PREDICT SYMBOL PREDICT SYMBOL PREDICT
...... 94.87544 - 108.50400
++++++ 108.50400 - 135.76112
000000 135.76112 - 163.01824
163.01824 - 190.27536
190.27536 - 217.53248
217.53248 - 231.16104
NOTE: 732 OBS HIDDEN

6. Special Models

6.1 INTRODUCTION

In Chapter 5 you were introduced to polynomial models that are used to analyze relationships that cannot be described by straight lines. Although polynomial models are very useful, they cannot provide adequate descriptions for all types of relationships. For example, in the data on fish growth, the polynomial provided a statistically significant fit, but the plot of the resulting curve, even for the fourth order polynomial, showed some features that did not fit the desired characteristics for such a curve. A number of other regression methods are available for such cases. The following methods are discussed in this chapter:

Log-linear model
: uses logarithms to provide a multiplicative model.

Strictly nonlinear model
: allows the fitting of virtually any smooth and connected functional relationship.

Spline function
: allows different functions, usually polynomial models, to be fitted for different regions of the data.

Indicator variables
: allow the estimation of shifts in the slope for various conditions.

Most of these techniques are more difficult to use than strictly linear or polynomial models. Furthermore, these techniques do not necessarily provide you with a more suitable model, but they can be useful in some cases. You invoke these techniques in the SAS System by either using special features and options in the standard linear models procedures or by using other specifically tailored procedures.

6.2 LOG-LINEAR (MULTIPLICATIVE) MODELS

A linear regression model using the logarithms of the variables is equivalent to estimating a multiplicative model.*

The log-linear model

$$\log(y) = \beta_0 + \beta_1(\log(x_1)) + \beta_2(\log(x_2)) + \ldots + \beta_m(\log(x_m)) + \varepsilon$$

---

* Either base 10 or base e can be used with identical results except for a change in the definition of the intercept. Base e is used in all examples.

is equivalent to the following model:

$$y = (e^{\beta_0})(x_1^{\beta_1})(x_2^{\beta_2}) \ldots (x_m^{\beta_m})(e^{\varepsilon}) \quad .$$

In this multiplicative model, the coefficients are elasticities. In other words, they measure the percent change in the dependent variable associated with a one percent change in the corresponding independent variable, holding constant all other variables. The intercept is a scaling factor. The error component in this model is also multiplicative and exhibits variation that is proportional to the magnitude of the dependent variable.

The log-linear model is illustrated by an example in which the weight of lumber in trees is to be estimated from external measurements of the trees. The dependent variable is WEIGHT, the weight of lumber in the tree. The independent variables are as follows:

HEIGHT is the height of the tree.

DBH is the diameter at breast height (approximately four feet).

AGE is the age of the tree.

GRAV is a measure of the specific gravity of the tree.

Obviously, the first two measurements are relatively easy to make, and a model for predicting tree weights using these variables could provide a low-cost estimate of timber yield. The other two variables are included to see if their addition to the model provides for better prediction. The data are shown in **Output 6.1**.

Output 6.1 Data Used to Estimate Tree Weights

OBS	DBH	HEIGHT	AGE	GRAV	WEIGHT
1	5.7	34	10	0.409	174
2	8.1	68	17	0.501	745
3	8.3	70	17	0.445	814
4	7.0	54	17	0.442	408
5	6.2	37	12	0.353	226
6	11.4	79	27	0.429	1675
7	11.6	70	26	0.497	1491
8	4.5	37	12	0.380	121
9	3.5	32	15	0.420	58
10	6.2	45	15	0.449	278
11	5.7	48	20	0.471	220
12	6.0	57	20	0.447	342
13	5.6	40	20	0.439	209
14	4.0	44	27	0.394	84
15	6.7	52	21	0.422	313
16	4.0	38	27	0.496	60
17	12.1	74	27	0.476	1692
18	4.5	37	12	0.382	74
19	8.6	60	23	0.502	515
20	9.3	63	18	0.458	766
21	6.5	57	18	0.474	345
22	5.6	46	12	0.413	210
23	4.3	41	12	0.382	100
24	4.5	42	12	0.457	122
25	7.7	64	19	0.478	539
26	8.8	70	22	0.496	815
27	5.0	53	23	0.485	194
28	5.4	61	23	0.488	280
29	6.0	56	23	0.435	296
30	7.4	52	14	0.474	462
31	5.6	48	19	0.441	200
32	5.5	50	19	0.506	229
33	4.3	50	19	0.410	125

(continued on next page)

(continued from previous page)

```
34    4.2    31    10    0.412     84
35    3.7    27    10    0.418     70
36    6.1    39    10    0.470    224
37    3.9    35    19    0.426     99
38    5.2    48    13    0.436    200
39    5.6    47    13    0.472    214
40    7.8    69    13    0.470    712
41    6.1    49    13    0.464    297
42    6.1    44    13    0.450    238
43    4.0    34    13    0.424     89
44    4.0    38    13    0.407     76
45    8.0    61    13    0.508    614
46    5.2    47    13    0.432    194
47    3.7    33    13    0.389     66
```

You first create a linear regression model and make provisions for a residual plot. The required SAS statements are as follows:

```
PROC REG;
   MODEL WEIGHT=HEIGHT DBH AGE GRAV;
   OUTPUT OUT=A P=PW R=RW;
PROC PLOT;
   PLOT RW*PW / VREF=0;
```

The results from PROC REG are shown in **Output 6.2**.

Output 6.2 Using PROC REG to Estimate Tree Weights

```
DEP VARIABLE: WEIGHT
                              ANALYSIS OF VARIANCE

                               SUM OF         MEAN
         SOURCE     DF        SQUARES       SQUARE       F VALUE       PROB>F

         MODEL       4     6570094.59   1642523.65       124.059       0.0001
         ERROR      42      556075.96  13239.90379
         C TOTAL    46     7126170.55

             ROOT MSE        115.0648     R-SQUARE        0.9220
             DEP MEAN        369.3404     ADJ R-SQ        0.9145
             C.V.            31.15413

                              PARAMETER ESTIMATES

                      PARAMETER         STANDARD      T FOR H0:
   VARIABLE  DF        ESTIMATE            ERROR    PARAMETER=0      PROB > |T|

   INTERCEP   1      -379.24822        206.69095         -1.835          0.0736
   HEIGHT     1      1.90015543       3.01673726          0.630          0.5322
   DBH        1       170.22033      16.23793314         10.483          0.0001
   AGE        1      8.14583480       4.02036310          2.026          0.0491
   GRAV       1     -1192.86848        548.92715         -2.173          0.0355
```

The model does fit rather well, and the coefficients have the expected signs. Surprisingly, the HEIGHT coefficient is not statistically significant. Furthermore, the significance of both the AGE and GRAV coefficients ($\alpha=0.05$) indicates that the model using the two external measurements is not adequate. Finally, the residual plot (**Output 6.3**) shows a pattern that suggests a poorly specified model.

Output 6.3 A Plot of Residuals

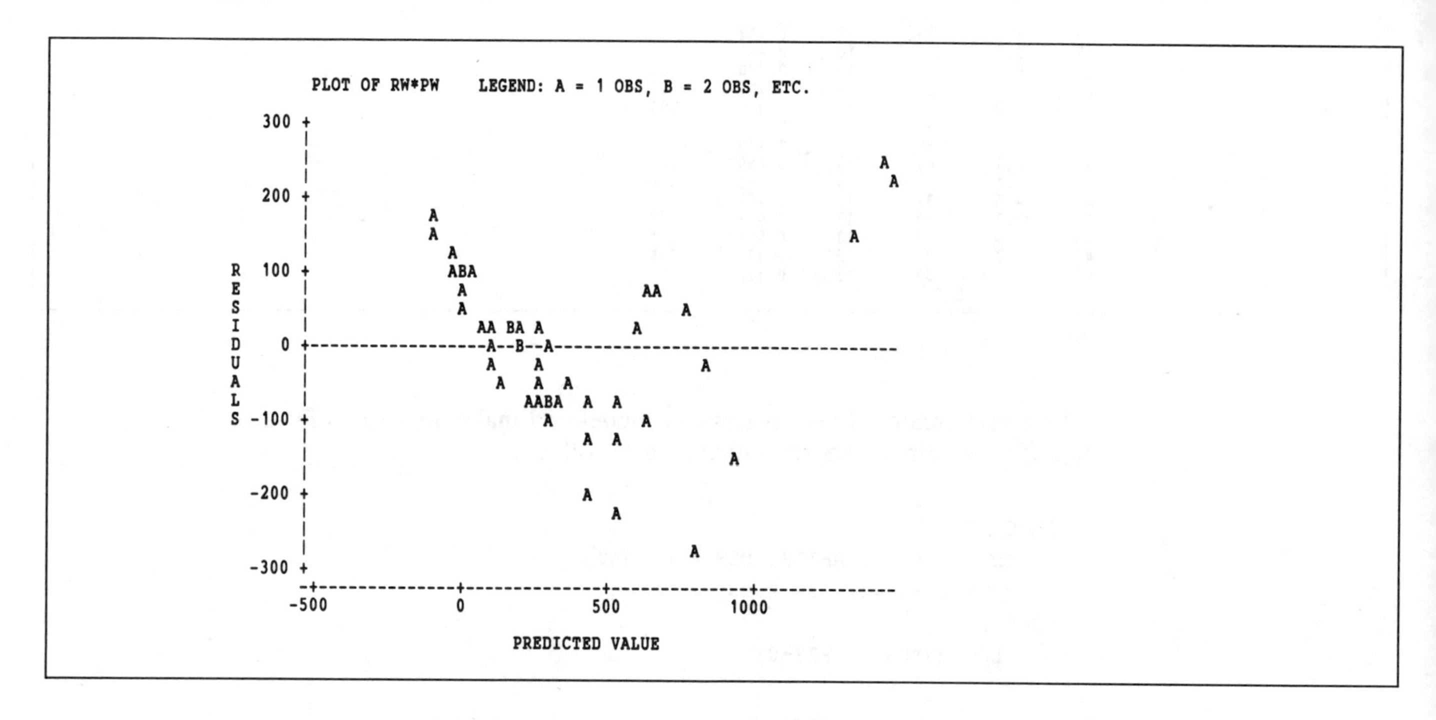

The pattern indicates the need for a curvilinear form, such as a quadratic response function. In this case, a multiplicative model is more suitable. The amount of lumber in a tree is a function of the volume of the trunk, which is in the shape of a cylinder. The volume of a cylinder is $\pi r^2 h$, where r is the radius and h is the height of the cylinder. For this reason, a multiplicative model for volume or weight using radius (or diameter) and height is appropriate. It is also reasonable to expect that age and gravity have a relative, or multiplicative, effect. A log-linear model is used to account for this multiplicative effect.

The log-linear model is implemented by obtaining the logarithms of the observed values in the DATA step. You do this by adding the following steps after the INPUT statement:

```
ARRAY X {5} DBH--WEIGHT;
ARRAY L {5} LDBH LHEIGHT LAGE LGRAV LWEIGHT;
DO I=1 TO 5;
   L {I}=LOG(X {I});
END;
DROP I;
```

The new variables LDBH--LWEIGHT are now available to use with the log-linear model. You can implement the model with PROC REG and make provisions for the residual plot by creating a data set with PLW and RLW as the predicted and residual values for the log-linear model with the following statements:

```
PROC REG;
   MODEL LWEIGHT=LDBH LHEIGHT LAGE LGRAV;
   OUTPUT OUT=B P=PLW R=RLW;
PROC PLOT;
   PLOT RLW*PLW / VREF=0;
```

The output from PROC REG is shown in **Output 6.4**.

Output 6.4 A Log-Linear Model Using PROC REG

```
DEP VARIABLE: LWEIGHT
                                  ANALYSIS OF VARIANCE

                                   SUM OF          MEAN
              SOURCE     DF       SQUARES        SQUARE        F VALUE        PROB>F

              MODEL       4   36.58763050    9.14690763        572.856        0.0001
              ERROR      42    0.67062265    0.01596721
              C TOTAL    46   37.25825316

                  ROOT MSE      0.1263614      R-SQUARE         0.9820
                  DEP MEAN       5.494661      ADJ R-SQ         0.9803
                  C.V.           2.299713

                                  PARAMETER ESTIMATES

                             PARAMETER         STANDARD        T FOR H0:
         VARIABLE  DF         ESTIMATE            ERROR      PARAMETER=0        PROB > |T|

         INTERCEP   1      -1.55823289       0.56527317           -2.757            0.0086
         LDBH       1       2.14477767       0.11910546           18.007            0.0001
         LHEIGHT    1       0.97784584       0.16960799            5.765            0.0001
         LAGE       1      -0.15509211       0.08050407           -1.927            0.0608
         LGRAV      1       0.10774830       0.26746579            0.403            0.6891
```

Note that although R^2 values and F ratios are not strictly comparable, the overall fit of this model is much better than that of the linear model. Furthermore, both the LDBH and LHEIGHT coefficients are highly significant, while neither of the other variables contributes significantly at the $\alpha=0.05$ level. Finally, the coefficients for LDBH and LHEIGHT are quite close to two and unity, respectively. These values are the values you would expect with a model based on the cylindrical shape of the tree trunk. The plot of the residual values against the predicted values appears in **Output 6.5**.

Output 6.5 Residual Plot of a Log-Linear Model

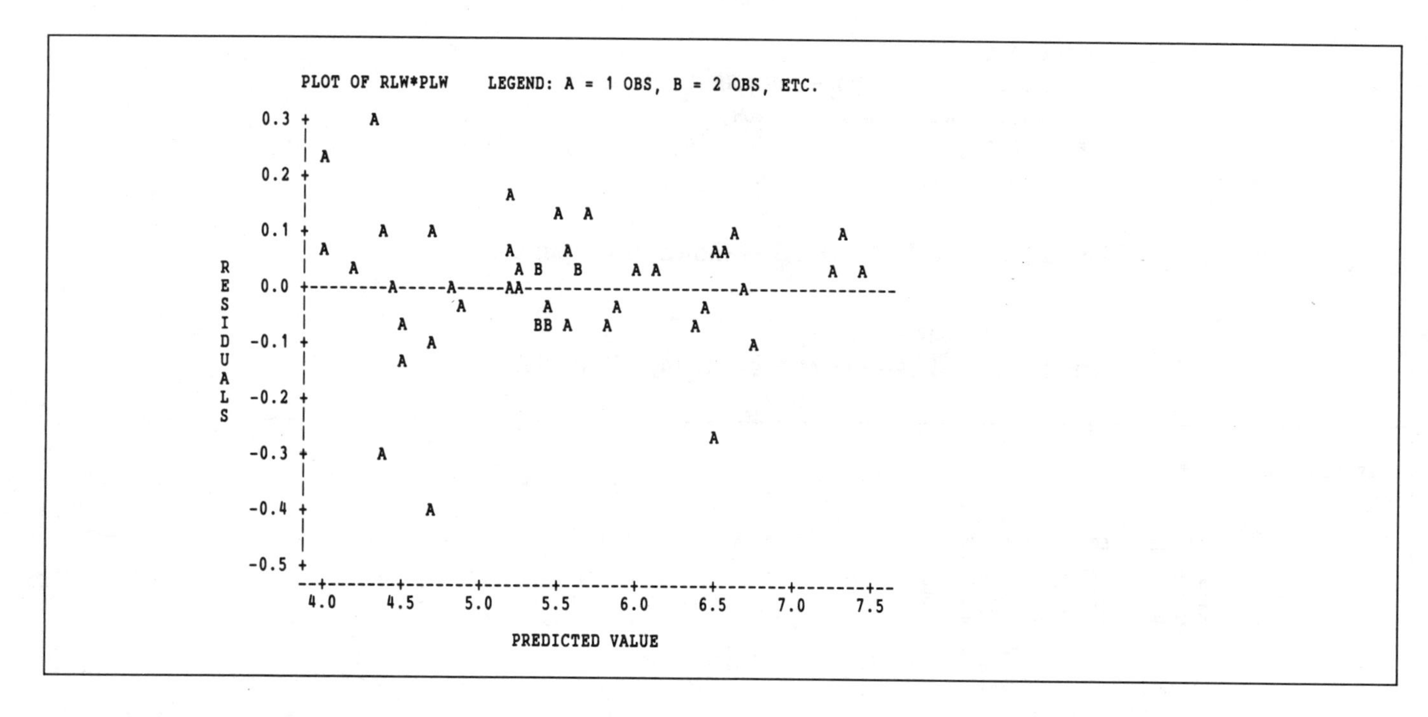

The residuals from this model (**Output 6.5**) appear to show virtually no patterns that suggest specification errors. However, there are some suspiciously large residuals for some of the smaller values of the predicted values that may require further scrutiny. Of course, these residuals show relative errors, and therefore, these residuals do not necessarily correspond to large absolute residuals. Additional procedures for examining residuals are presented in Chapter 3.

It may be of interest to see how well the log-linear model estimates the actual weights rather than the logarithms of the weights. To do this, you can take antilogs or exponentiate the predicted values from the log-linear model. This is done with a new DATA step:

```
DATA RESID;
   SET B;
   PMW=EXP(PLW);
   RMW=WEIGHT-PMW;
```

The variable PMW is the antilog of the predicted logarithm weights, and RMW is the residual. You can use PROC MEANS to obtain the mean and the corresponding sum of squares of these residuals. With PROC MEANS you obtain

mean residual=3.970
sum of squares=76,260 .

These statistics illustrate the fact that estimated, or predicted, values obtained in this manner are biased and, furthermore, that the resulting sum of squared residuals is larger than that obtained by the linear model. The plot of these residuals is shown in **Output 6.6**.

Output 6.6 Plot of Exponentiated Log Residuals

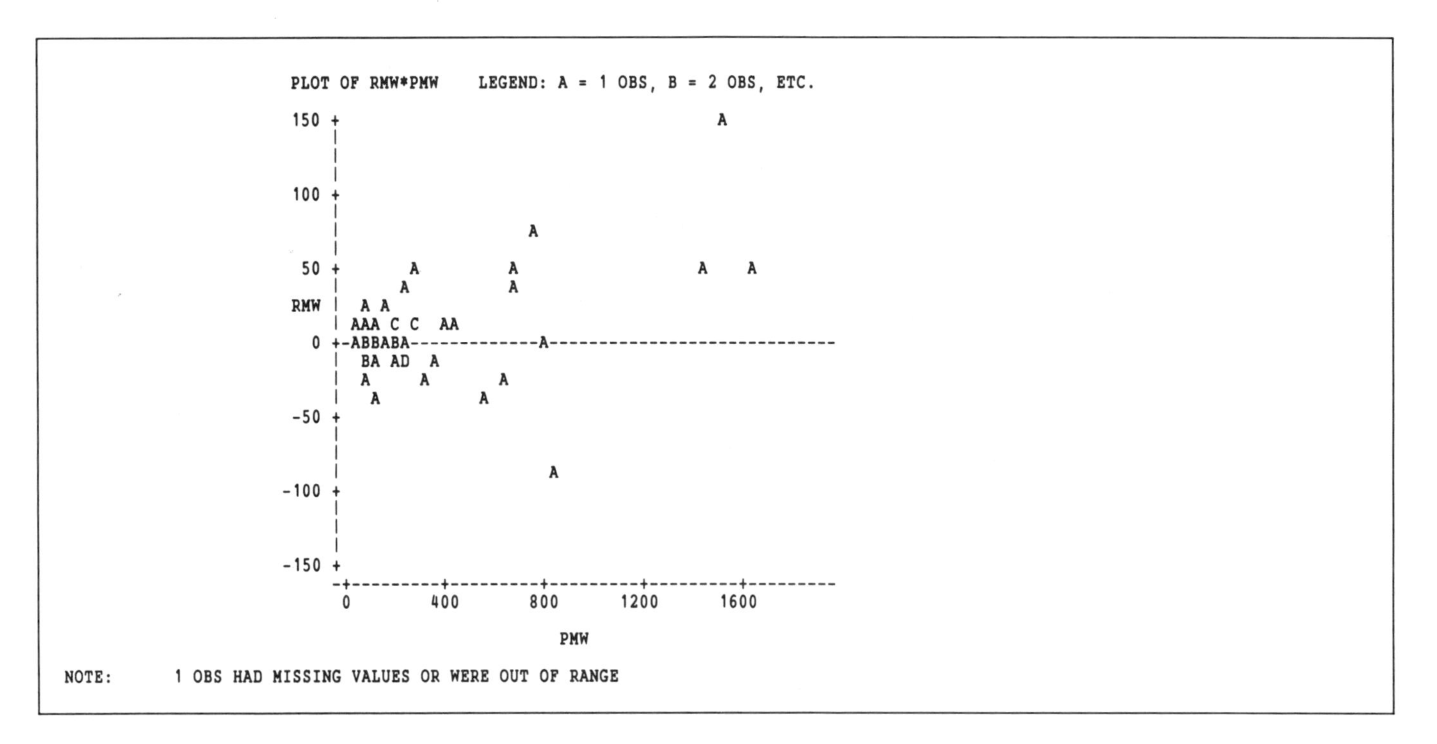

These residuals do not have the obvious specification error pattern exhibited by the residuals from the linear model. They do show the typical pattern of multiplicative errors, where larger residuals are associated with larger values of the response variable.

6.3 NONLINEAR MODELS

When a model is nonlinear in the parameters, the entire process of estimation and statistical inference is radically altered. This happens mainly because the normal equations that are solved to obtain least-squares parameter estimates are themselves nonlinear. Solutions of systems of nonlinear equations are not usually available in closed form but must be obtained by numerical methods. For this reason, closed-form expressions for the partitioning of the sums of squares and the consequently obtained statistics for making inferences about parameters in nonlinear models are also unavailable.

For most applications, the solutions to the normal equations are obtained by means of an iterative process. The process starts with some preliminary estimates of the parameters. These estimates are used to calculate a residual sum of squares and an indication of which modifications in these estimates may result in reducing the residual sum of squares. This process is repeated until it appears that no further modifications of the parameter estimates result in a reduction of the residual sum of squares.

Variances of the parameter estimates are computed using asymptotic approximations; that is, the variances are based on the assumption of large sample sizes.

Before continuing, it is important to note a few characteristics of nonlinear regression:

- There is no universally best or most efficient method for obtaining the solution of the normal equations.
- There is no guarantee that the solution obtained is the global least-squares estimate.
- The given variances of the estimated coefficients are only asymptotic approximations, assuming that the sample used to calculate the estimates was very large.

Nonlinear models are performed in the SAS System with PROC NLIN. PROC NLIN is illustrated here using a simple model. Comments on other applications are given in section **6.4**.

A popular and simple nonlinear model is the exponential decay curve that describes how radiation levels from a source decrease as the amount of radioactive material decays over time. The amount, or counts, of radiation should decay according to the following model:

$$\text{COUNT} = A + B^{-C(\text{TIME})} + \varepsilon$$

where

COUNT	is the measure of the amount of radiation.
$A+B$	is the initial level of radiation at time zero.
A	is the ultimate level of radiation at time ∞; usually, this is zero.
C	is a measure of the intensity of the radiation. It is related to the half-life of the radioactive substance by the formula $C=\ln(2)/\text{T2}$, where T2 is the half-life.
TIME	is the elapsed time.
ε	is the random error.

In this example, radiation counts were recorded nineteen times during a 400-hour time period. The data are given in **Output 6.7**. Three additional times have been added to the set of time periods to predict the estimated curve beyond the range of observed data.

Output 6.7 Data for an Exponential Decay Curve

TIME	COUNT	TIME	COUNT
0	383.0	138	261.0
14	373.0	165	244.0
43	348.0	224	200.0
61	328.0	236	197.0
69	324.0	253	185.0
74	317.0	265	180.0
86	307.0	404	120.5
90	302.0	434	112.5
92	298.0	500	.
117	280.0	750	.
133	268.0	1000	.

PROC NLIN is used to estimate the exponential decay curve for these data. As previously noted, the least-squares estimation procedure is iterative; hence, it requires some preliminary parameter estimates.

In this rather simple example, these initial estimates are not difficult to make. Radiation levels at (or near) time zero have been observed, so this value is used to specify the initial radiation value, $A+B=380$. For this particular substance, all radiation is expected to cease ultimately; hence, the asymptote, B, is zero. Thus, the initial estimate of the parameter is

$$B = 380 - A = 380 \quad .$$

Then use these values to see what value of the decay parameter, C, satisfies one (or more) typical data values. You do this by inserting the initial estimates into the equation for some arbitrary data value, say the observation for TIME=117, where COUNT=280. The resulting equation is

$$280 = 380e^{117(C)} \quad .$$

Solving for C gives an initial estimate of $C=-0.0026$. Now you can implement PROC NLIN:

```
PROC NLIN;
   PARMS A=0 B=380 C=-.0026;
   MODEL COUNT=A+B*EXP(C*TIME);
   OUTPUT OUT=A PREDICTED=PNLC RESIDUAL=RNLC;
```

The initial values of the parameters are specified in the PARMS (for parameters) statement, which is followed by the MODEL statement. You will notice that the MODEL statement in PROC NLIN is very different from the MODEL statement in PROC REG. This is due to the fact that for PROC REG the regression equation is known, whereas in PROC NLIN you must supply the specific formula. You can also request that a new data set be produced that includes predicted and residual values. In PROC NLIN this is implemented in the same manner as for PROC REG. The resulting output is given in **Output 6.8**.

Output 6.8 PROC NLIN Used with Radiation Data

NON-LINEAR LEAST SQUARES ITERATIVE PHASE

DEPENDENT VARIABLE: COUNT METHOD: DUD

ITERATION	A	B	C	RESIDUAL SS
-4	0	380.000000	-0.0026000000	978.430671942
-3	0.100000000	380.000000	-0.0026000000	986.802108059
-2	0	418.000000	-0.0026000000	15946.476133676
-1	0	380.000000	-0.0028600000	957.965147627
0	0	380.000000	-0.0028600000	957.965147627
1	-13.673237329	402.780002	-0.0027418916	133.972108448
2	-13.253715721	402.348133	-0.0027354739	131.362409159
3	-13.088616085	402.198371	-0.0027379025	131.362241927
4	-13.169073474	402.270520	-0.0027367041	131.359621852
5	-13.170998955	402.269527	-0.0027366423	131.359599126
6	-13.171723713	402.270197	-0.0027366351	131.359599043

NOTE: CONVERGENCE CRITERION MET.

```
                NON-LINEAR LEAST SQUARES SUMMARY STATISTICS     DEPENDENT VARIABLE COUNT

                  SOURCE                DF SUM OF SQUARES      MEAN SQUARE

                  REGRESSION             3   1444588.1404      481529.3801
                  RESIDUAL              16       131.3596          8.2100
                  UNCORRECTED TOTAL     19   1444719.5000

                  (CORRECTED TOTAL)     18    114151.9211

                  PARAMETER      ESTIMATE     ASYMPTOTIC              ASYMPTOTIC 95 %
                                              STD. ERROR           CONFIDENCE INTERVAL
                                                                   LOWER          UPPER
                  A           -13.1717237  12.433899133  -39.53028138    13.18683396
                  B           402.2701967  11.436640360  378.02572265   426.51467069
                  C            -0.0027366   0.000139647   -0.00303267    -0.00244060


                           ASYMPTOTIC CORRELATION MATRIX OF THE PARAMETERS

                    CORR                A                 B                 C

                    A              1.0000           -0.9923           -0.9804
                    B             -0.9923            1.0000            0.9544
                    C             -0.9804            0.9544            1.0000

NOTE: ALL ASYMPTOTIC STATISTICS ARE APPROXIMATE. REFERENCE: RALSTON AND JENNRICH, TECHNOMETRICS, FEBRUARY 1978, P 7-14.
```

The first portion of the output provides information on the iterative phase for obtaining the solution of the normal equations. The notation METHOD: DUD refers to the specific iterative method used. In this case the (default) *derivative free* method was implemented (see section **6.4**). This method checks a selected set of preliminary parameter estimates in the neighborhood of the values specified in the PARMS statement. Parameter estimates are used to choose the direction of further iterations. In this case, the first estimates are quite close to the final obtained values; hence, the iterations converge quickly to the final estimates.

The second portion of the output gives the partitioning of the sum of squares that is calculated directly from the actual and predicted values. Since most nonlinear models do not contain an identifiable intercept term, the uncorrected sum of squares is used as the total sum of squares. However, the corrected total sum of squares is also provided so that an R^2 value corresponding to the R^2 value used in linear models may be calculated. In this example, that value is $R^2=1-(SSE/CSS)$, or 0.9988. You should be aware that the choice of an R^2 statistic with nonlinear regression is not as straightforward as in the case of linear regression. See Kvalseth (1985) for details.

The final section of the output gives the parameter estimates and other statistics useful for making inferences on these parameters. These statistics show that the estimated asymptote, B, is not zero. However, the 95% confidence interval does include that value, so you may conclude that the model conforms to theory. The decay constant can be used to estimate the half-life, $T2=\ln(2)/c=202$ hours for this example. Finally, you can see that the coefficient estimates are highly correlated with each other. This is an often encountered phenomenon in nonlinear regression. The output finally states that the standard errors are asymptotic approximations.

You can use the predicted and residual values in data set A to make plots. The inclusion of TIME values beyond the range of those observed provides information on how the estimated decay curve looks in that area. The plot of predicted values is produced by the following statements:

```
PROC PLOT;
   PLOT PNLC*TIME;
```

The plot is shown in **Output 6.9**. You can see that the radiation still continues after 1000 hours.

Output 6.9 Plot of Predicted Values

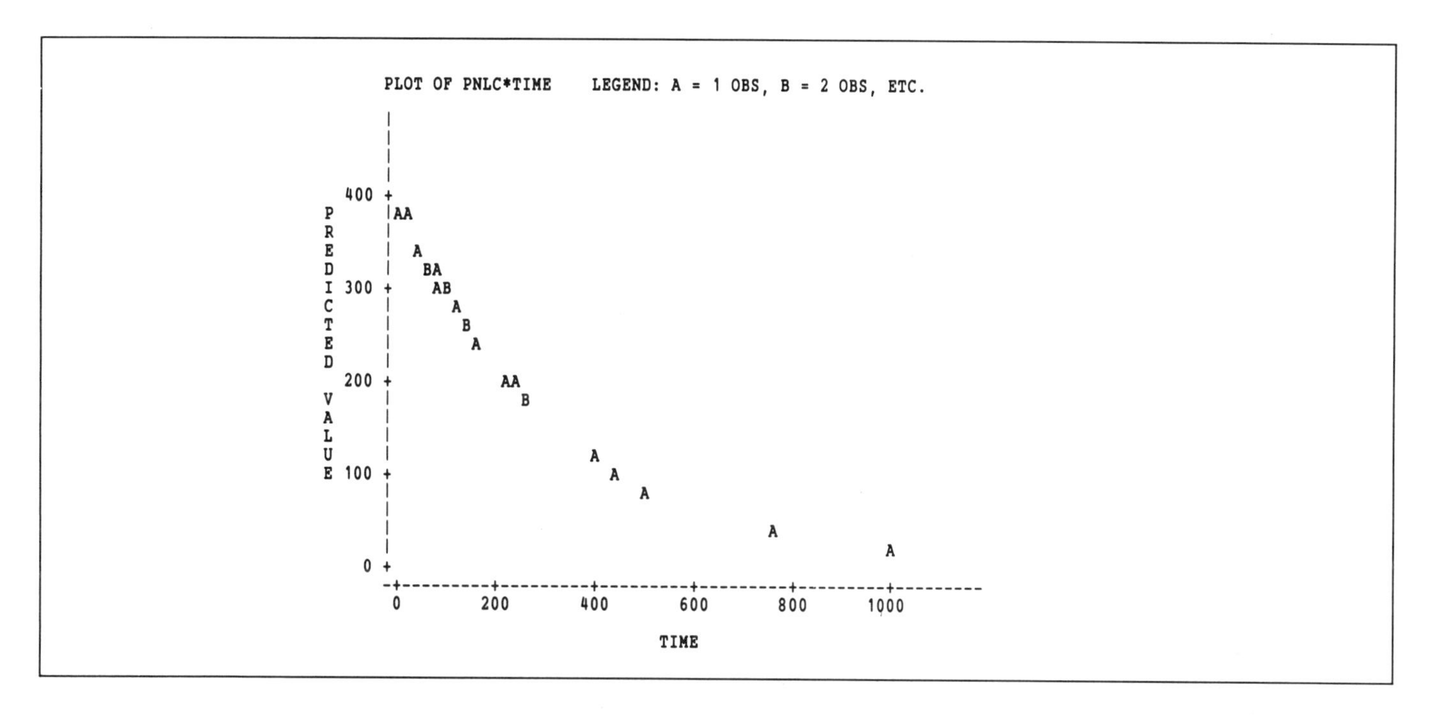

You do not need these extrapolation points for the residual plot. You can first create a new data set without them and then plot:

```
DATA B;
   SET A;
   IF TIME<450;
PROC PLOT;
   PLOT RNLC*TIME / VREF=0;
```

The plot is given in **Output 6.10**.

Output 6.10 Residual Plot

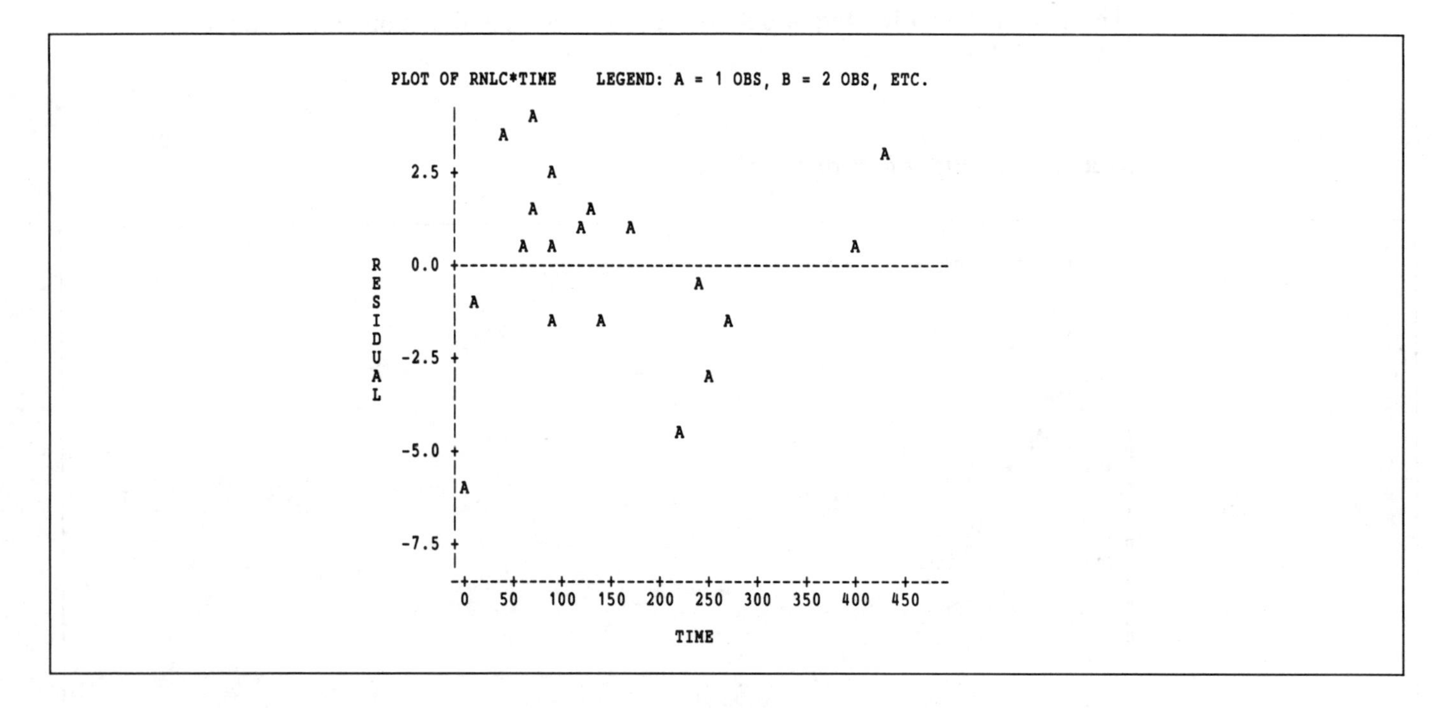

The residual plot exhibits some suspicious patterns that suggest the possibility of other sources of radiation.

It is interesting to compare the results of fitting the theoretically correct exponential model with the results obtained by using a quadratic polynomial model. The results for the polynomial model obtained using PROC REG are shown in **Output 6.11**. TSQ is defined as TIME*TIME in the DATA step. The required SAS statements are as follows:

```
DATA RADIAT1;
   SET RADIAT;
   TSQ=TIME*TIME;
PROC REG DATA=RADIAT1;
   MODEL COUNT=TIME TSQ / SS1 SS2;
   OUTPUT OUT=A P=PRAD;
```

Output 6.11 Quadratic Model Results

```
DEP VARIABLE: COUNT
                              ANALYSIS OF VARIANCE

                         SUM OF         MEAN
        SOURCE    DF    SQUARES       SQUARE      F VALUE      PROB>F

        MODEL      2  114086.66  57043.33191    13986.088      0.0001
        ERROR     16 65.25722837  4.07857677
        C TOTAL   18  114151.92

          ROOT MSE     2.019549     R-SQUARE       0.9994
          DEP MEAN     264.6316     ADJ R-SQ       0.9994
          C.V.        0.7631548
```

(continued on next page)

(continued from previous page)

PARAMETER ESTIMATES

VARIABLE	DF	PARAMETER ESTIMATE	STANDARD ERROR	T FOR H0: PARAMETER=0	PROB > |T|	TYPE I SS	TYPE II SS
INTERCEP	1	386.91763	1.11748182	346.241	0.0001	1330567.58	488950.33
TIME	1	-1.01987814	0.01341873	-76.004	0.0001	110575.51	23560.38153
TSQ	1	0.000891277	0.000030377	29.341	0.0001	3511.15030	3511.15030

It is seen that the quadratic model actually fits the data better with a residual mean square of 4.08 compared to 8.21 for the exponential model. Since the polynomial model is easier to implement than a nonlinear model, the polynomial model would appear to be the model you should use. This is not necessarily the correct conclusion.

One reason for this can be seen in the plot of predicted values obtained by the following SAS statements:

```
PROC PLOT DATA=A;
   PLOT PRAD*TIME / HPOS=60 VPOS=25;
```

The plot, given in **Output 6.12**, clearly shows the well-known failure of polynomial models for extrapolation. In this case the predicted values show increased radiation when TIME exceeds 600, which is a physical impossibility.

Output 6.12 Plot of Predicted Values for the Quadratic Model

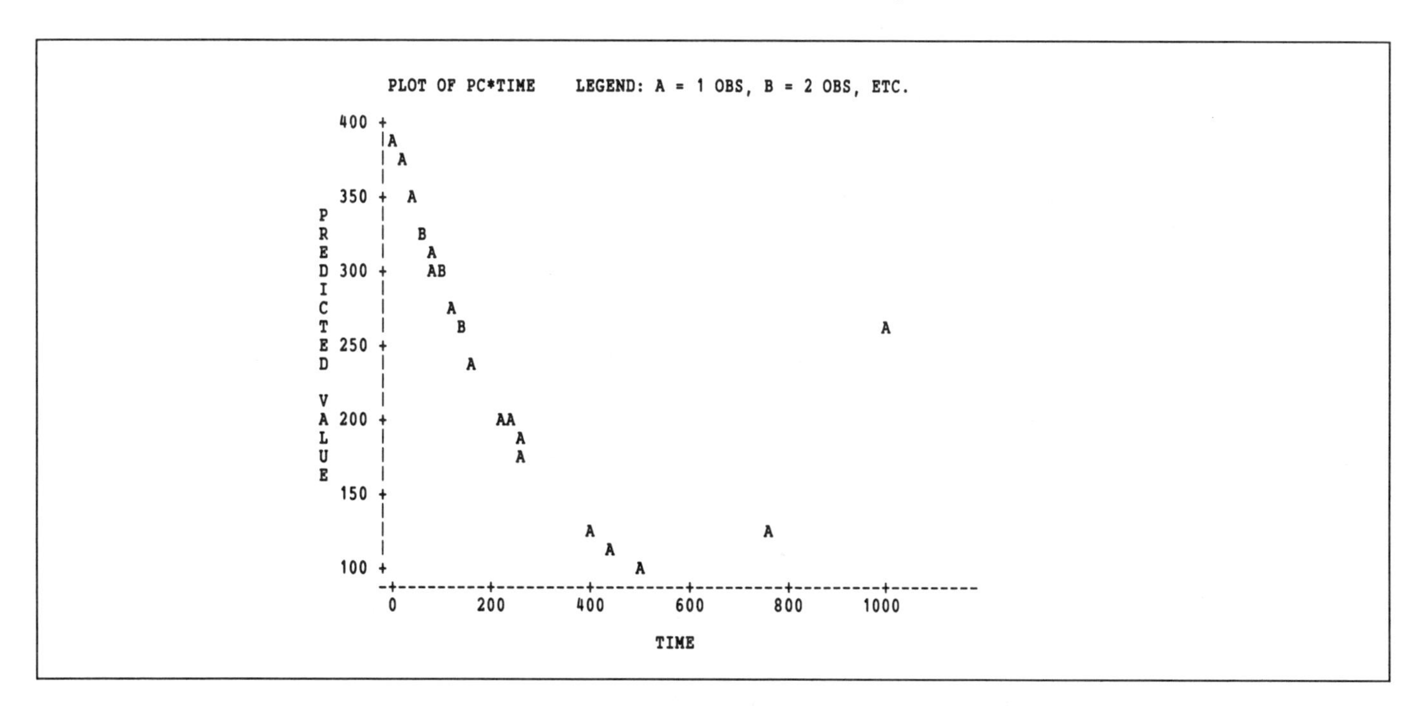

In addition, the estimated coefficients of the polynomial equation

$$\text{COUNT} = 388.92 - 1.0199(\text{TIME}) + 0.0008913(\text{TIME})^2$$

have, with the exception of the intercept, no meaning, whereas in the exponential model all coefficients are relevant to the physical phenomenon. In other words, the polynomial model may be useful for curve fitting (or smoothing) but may not provide a useful model for explaining a physical phenomenon.

6.4 ADDITIONAL COMMENTS ON PROC NLIN

The example of nonlinear regression presented in the previous section used a very simple model and was easily analyzed using PROC NLIN. Unfortunately, the difficulty with the estimation process increases much more rapidly for nonlinear regression models than it does for linear models as you increase the number of parameters. For this reason, PROC NLIN provides a number of features to assist in providing estimates and other inferences that may be useful for more complicated models. The following paragraphs summarize these features.

Estimation In the previous example, the default derivative free (DUD) method was used to find the solution to the normal equations. Although convenient to use, it is not usually the most efficient method. PROC NLIN provides three alternative iterative methods that can be implemented by applying the appropriate PROC method option and by supplying the partial derivatives of the model function with respect to the parameters.

Initial values The initial values of the parameters required for starting the iterative solution process are not always easily obtained. The NLIN procedure allows for the implementation of a grid search procedure that computes the residual sum of squares for a grid of potential parameter values. Then, the grid search procedure starts iterations at the combination of values that gives the smallest residual sum of squares. Plots of the results of the grid search may be requested as a PROC option.

Convergence Occasionally, the output of the iterative process indicates that convergence criteria have not been met. This means that a satisfactory least-squares solution has not been found. Since this is usually the result of a poor choice of initial values, a grid search should be implemented, or if one has already been done, those results should be used to determine the bounds for a new grid search. Occasionally, using a different solution procedure may help.

Local minima The solution found by the iterative procedure may not be the correct or global minimum residual sum of squares solution. Such an occurrence may result in a residual sum of squares that is obviously too large, or this may yield unreasonable parameter estimates. However, this is not always the case. Therefore, if convergence to a nonglobal least-squares estimate is suspected, you may want to implement a polynomial regression to see if the nonlinear model produces a much larger error mean square. Again, different starting values or estimation methods should be used if the estimates are in doubt.

Restrictions on parameters A BOUNDS statement may be used to restrict parameter estimates. However, the use of BOUNDS statements may invite finding local optima.

Derivative specification Programming statements such as those used in the DATA step can be implemented between the PARMS and MODEL statements. Such statements can be used to simplify expressions and save computing time. For example, derivatives of exponential functions involve the same exponentials. Defining an exponential function once in a program step allows its repeated use without recomputation.

Specific instructions for implementing these various alternatives and options are given in the *SAS User's Guide: Statistics, Version 5 Edition.*

6.5 SPLINE MODELS

In a review paper on spline models, Smith (1979) gives the following definition:

> Splines are generally defined to be piecewise polynomials of degree n whose function values and first $n-1$ derivatives agree at points where they join. The abscissas of these joint points are called knots. Polynomials may be considered a special case of splines with no knots, and piecewise (sometimes also called grafted or segmented polynomials) with fewer than the maximum number of continuity restrictions may also be considered splines. The number and degrees of polynomial pieces and the number and position of knots may vary in different situations.

In this section an example of a spline function of degree two with only one knot is presented. First, assume that the position of the knot is known, and then consider the case where the position of the knot must be estimated.

The "+" function facilitates the implementation of spline regressions. The "+" function is defined for a variable x as follows:

$$x^+ = x \quad \text{if } x>0$$
$$x^+ = 0 \quad \text{if } x<0 \quad .$$

Thus, the spline function with quadratic polynomials and the single knot at $x=t$ is defined as follows:

$$y = \beta_0 + \beta_1 x + \beta_2 x_2 + \beta_3 (x - t)^+ + \beta_4((x - t)^+)^2 \quad .$$

This spline function is illustrated using the fish growth data for TEMP=29 used in Chapter 5 (**Output 5.1**). These data appear to indicate an abrupt change in the growth of fish at about eighty days; hence, you assume the knot to be at that point. This model is readily illustrated as a linear regression model by computing the required "+" function values in the DATA step as follows:

```
AL1=MAX (AGE-80,0);
AS1=AL1*AL1;
```

The MAX function returns the maximum value of the two values specified in the argument. Thus, AL1 and AS1 become the "+" function values for AGE and AGE^2. The resulting data are presented in **Output 6.13**.

Output 6.13 Data for Spline

OBS	AGE	LENGTH	ASQ	AL1	AS1
1	14	590	196	0	0
2	21	910	441	0	0
3	28	1305	784	0	0
4	35	1730	1225	0	0
5	42	2140	1764	0	0
6	49	2725	2401	0	0
7	56	2890	3136	0	0
8	63	3685	3969	0	0
9	70	3920	4900	0	0
10	77	4325	5929	0	0
11	84	4410	7056	4	16
12	91	4485	8281	11	121
13	98	4515	9604	18	324
14	105	4480	11025	25	625
15	112	4520	12544	32	1024
16	119	4545	14161	39	1521
17	126	4525	15876	46	2116
18	133	4560	17689	53	2809
19	140	4565	19600	60	3600
20	147	4626	21609	67	4489
21	154	4566	23716	74	5476

You can produce the spline model with PROC REG using the following statements:

```
PROC REG;
   MODEL LENGTH=AGE ASQ AL1 AS1 / SS1;
   OUTPUT OUT=SPL1 P=PSPL1 R=RSPL1;
```

The results are given in **Output 6.14**.

You can immediately see that the residual mean square is less than that for the fourth degree polynomial (**Output 3.2**). Furthermore, this spline function does not exhibit the peculiar hook estimated by the fourth degree polynomial model at the later ages. As in ordinary polynomial regression, the Type I sums of squares are useful. In this example, the Type I sums of squares show that the quadratic "+" function coefficient (AS1) is not significant. In other words, the quadratic component of the model does not appear to change at the knot. The overall quadratic coefficient also does not appear to contribute to the model, but a definitive statement on this coefficient can only be made if the model is reestimated without the quadratic "+" function term. Implementation of such a model (not shown) does indeed show that the simple quadratic component is not needed. The model using only AGE and AL1 produces the following equation:

$$\text{LENGTH} = -327.62 + 60.175(\text{AGE}) - 58.863(\text{AGE} - 80)^{+} \quad .$$

The residual mean square for this model, 6209, is smaller than that for the full-spline model. The coefficients show a growth rate of 60.175 until day eighty and a rate of 1.312(60.175−58.863) after that time. You can test if that growth rate is significantly different from zero by adding the following TEST statement to PROC REG:

```
AFTER80: TEST AGE+AL1=0;
```

Output 6.14 Spline Regression Results

```
DEP VARIABLE: LENGTH
                              ANALYSIS OF VARIANCE

                           SUM OF          MEAN
        SOURCE      DF    SQUARES        SQUARE       F VALUE        PROB>F

        MODEL        4  38352134.53   9588033.63     1395.963        0.0001
        ERROR       16    109894.43   6868.40165
        C TOTAL     20  38462028.95

            ROOT MSE      82.87582     R-SQUARE        0.9971
            DEP MEAN      3524.619     ADJ R-SQ        0.9964
            C.V.          2.351341

                              PARAMETER ESTIMATES

                      PARAMETER         STANDARD      T FOR H0:
  VARIABLE  DF         ESTIMATE            ERROR    PARAMETER=0       PROB > |T|

  INTERCEP   1       -320.99363        130.73411         -2.455           0.0259
  AGE        1      59.63239312       6.13763131          9.716           0.0001
  ASQ        1      0.008320235       0.06266510          0.133           0.8960
  AL1        1     -61.42212183       6.99669993         -8.779           0.0001
  AS1        1       0.01670962       0.06230962          0.268           0.7920

  VARIABLE  DF        TYPE I SS

  INTERCEP   1        260881728
  AGE        1      30305600.11
  ASQ        1       7484972.88
  AL1        1        561067.60
  AS1        1        493.94493
```

The results of this command are printed after the coefficient estimates. The results for this example are as follows:

```
TEST: AFTER80    NUMERATOR:     14677.3  DF:   1  F VALUE:  2.3640
                 DENOMINATOR:   6208.69  DF:  18  PROB >F : 0.1416
```

Thus, a growth rate cannot be established beyond eighty days according to this model.* In other words, this model suggests that growth stops after eighty days.

Estimating a spline model where the position of the knot itself must be estimated is a nonlinear regression problem. You can estimate such a model using PROC NLIN. Since the model with the knot at eighty days does fit rather well, use the estimates from this model as initial values. You can also again use the default DUD method, especially since the derivative with respect to the knot is somewhat difficult to define. The required statements are as follows:

```
PROC NLIN;
   PARMS B0=-321
         B1=60
         B2=.0083
         B3=-61
         B4=.017
         KNOT=80;
   AL1=MAX (X-KNOT,0);
   AS1=AL1*AL1;
   MODEL LENGTH=B0+B1*AGE+B2*ASQ+B3*AL1+B4*AS1;
   OUTPUT OUT=C PREDICTED=PSPL RESIDUAL=RSPL;
```

* The estimate and standard error of that coefficient may be obtained with an ESTIMATE statement in PROC GLM.

Note that the "+" function values are calculated in the PROC step rather than in the DATA step since the values of this variable change as iterations converge to the least-squares solution. The results of the NLIN procedure are given in **Output 6.15**. The portion of the output describing the iterative process is not reproduced.

Output 6.15 Spline Regression: Unknown Knot

```
  NON-LINEAR LEAST SQUARES SUMMARY STATISTICS        DEPENDENT VARIABLE LENGTH

       SOURCE                 DF SUM OF SQUARES      MEAN SQUARE

       REGRESSION              6   299248628.07      49874771.34
       RESIDUAL               15       95128.93          6341.93
       UNCORRECTED TOTAL      21   299343757.00

       (CORRECTED TOTAL)      20    38462028.95

       PARAMETER       ESTIMATE     ASYMPTOTIC             ASYMPTOTIC 95 %
                                    STD. ERROR          CONFIDENCE INTERVAL
                                                         LOWER          UPPER
       B0         -253.0000000  133.35361141 -537.23541385  31.235413846
       B1           55.2597403    6.56131962   41.27467170  69.244808821
       B2            0.0649351    0.07078315   -0.08593508   0.215805208
       B3          -61.4335538    6.72294776  -75.76312327 -47.103984317
       B4           -0.0856049    0.08997035   -0.27737141   0.106161668
       KNOT         77.3774477    1.74129851   73.66597189  81.088923503


               ASYMPTOTIC CORRELATION MATRIX OF THE PARAMETERS

 CORR          B0           B1           B2           B3           B4         KNOT

 B0        1.0000      -0.9495       0.8842      -0.5029      -0.6960      -0.2391
 B1       -0.9495       1.0000      -0.9816       0.6086       0.7727       0.3221
 B2        0.8842      -0.9816       1.0000      -0.6530      -0.7872      -0.4009
 B3       -0.5029       0.6086      -0.6530       1.0000       0.0727      -0.2166
 B4       -0.6960       0.7727      -0.7872       0.0727       1.0000       0.7130
 KNOT     -0.2391       0.3221      -0.4009      -0.2166       0.7130       1.0000

NOTE: ALL ASYMPTOTIC STATISTICS ARE APPROXIMATE. REFERENCE: RALSTON AND
      JENNRICH, TECHNOMETRICS, FEBRUARY 1978, P 7-14.
```

The estimates produced by this procedure are, indeed, quite similar to those obtained when the knot was assumed to be eighty days. As in the case of the sequential polynomial plot, the resolution of a printer plot is insufficient to show the relationship of the actual and predicted values. A plot produced by SAS/GRAPH software shows more detail. The following SAS statements are required:

```
PROC GPLOT;
   PLOT LENGTH*AGE=1 PSPL1*AGE=2 / OVERLAY;
   SYMBOL1 V=SQUARE I=JOIN C=BLACK;
   SYMBOL2 V=S I=SPLINE C=BLACK;
```

This plot is shown in **Output 6.16**.

Output 6.16 SAS/GRAPH Software Plot Showing Actual and Predicted Values

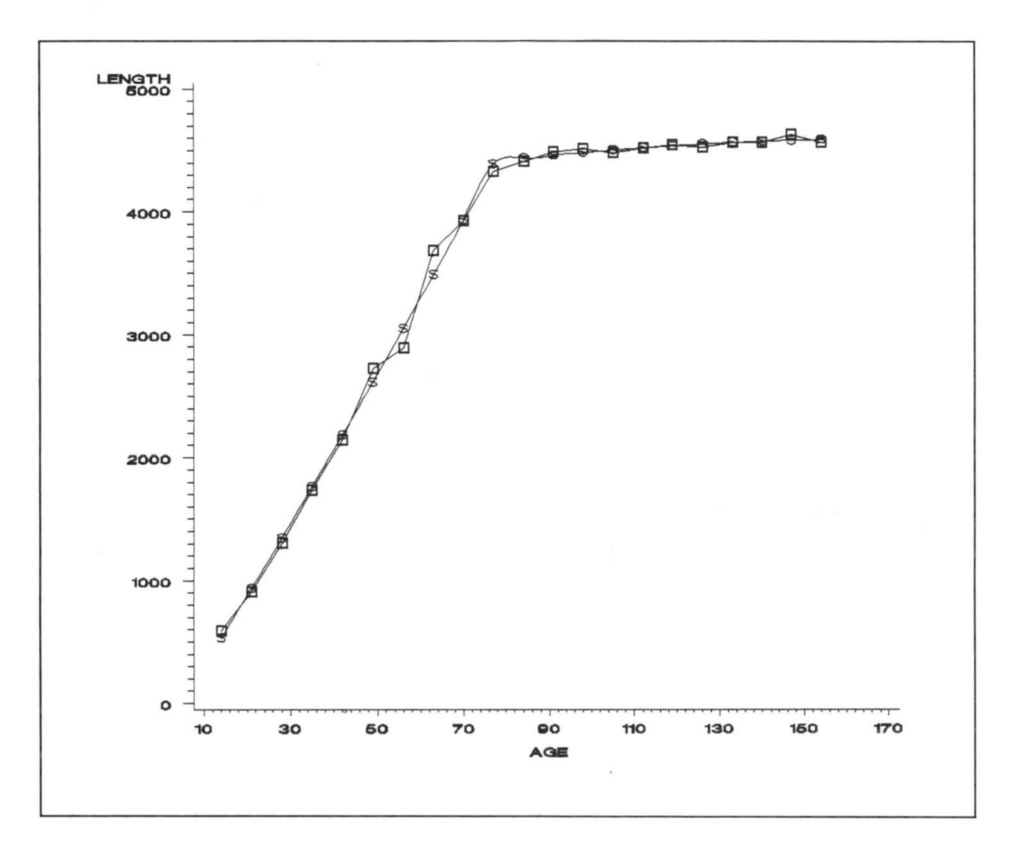

The following SAS statements are required for the residual plot for the spline regression:

```
PROC PLOT;
   PLOT RSPL1*AGE / HPOS=40 VPOS=25 VREF=0;
```

This plot is shown in **Output 6.17**.

These residuals show less of a pattern to suggest an inadequate model than do the residuals for the other polynomial models. There are still some rather large residuals at about forty to sixty days, but their pattern does not indicate anything that can be fitted by a reasonably smooth curve.

Again, it must be emphasized that a relatively simple example was used. Spline regression analysis may become quite difficult, particularly if multiple unknown knots and various continuity restrictions are required. Estimation of parameters can become especially vexing if the knots are somewhat indistinct or if more than the true number of knots is specified.

Output 6.17 Residual Plot: Spline Regression with Unknown Knot

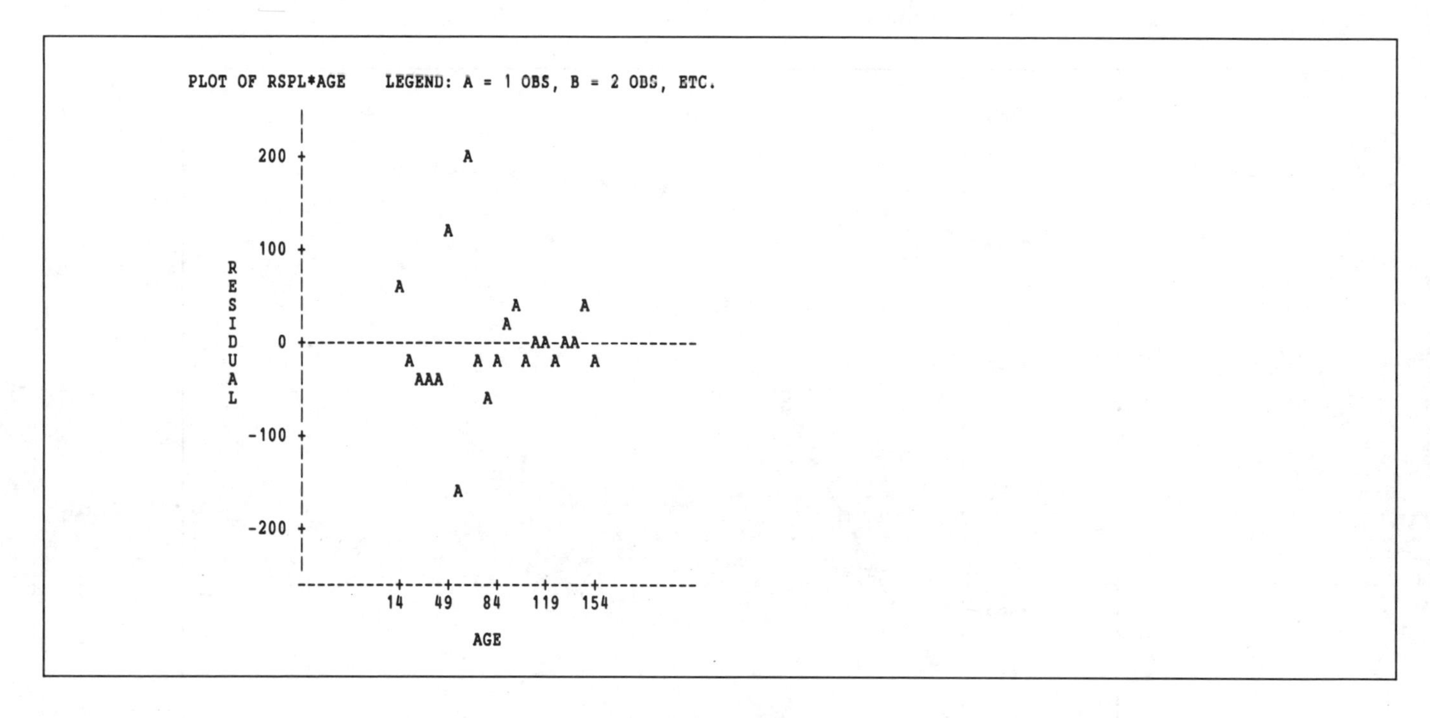

6.6 INDICATOR VARIABLES

The independent variables in a regression model are usually quantitative; that is, they have a defined scale of measurement. Occasionally, you may need to include qualitative, or categorical, variables in such a model. You can accomplish this by using indicator, or dummy, variables.

The example in Chapter 2 concerned a model for determining operating costs of airlines. It has been argued that long-haul airlines have lower operating costs than do short-haul airlines. Use the variable ASL, the average stage length, to define a variable TYPE as follows:

```
IF ASL<1200 THEN TYPE=0;
ELSE TYPE=1;
```

The TYPE variable is a dummy, or indicator, variable that classifies airlines into two groups or classes. TYPE=0 defines the short-haul lines with average stage lengths of less than 1200 miles, and TYPE=1 defines the long-haul lines that have average stage lengths of 1200 miles or longer.*

Implement the regression with the following statements:

```
PROC REG;
   MODEL CPM=UTL SPA ALF TYPE;
```

When TYPE=0, INTERCEP is the intercept for the model describing the short-haul lines. When TYPE=1, the intercept is the sum of the INTERCEP and TYPE

* This is an arbitrary definition used for this example.

coefficients describing the long-haul lines. Since the other coefficients retain their original interpretation, the TYPE coefficient simply estimates the difference in levels of operating costs between the two types of airlines. A negative coefficient indicates lower costs for the long-haul lines. The results are shown in **Output 6.18**.

Output 6.18 Using an Indicator Variable for an Intercept Shift

```
DEP VARIABLE: CPM
                                  ANALYSIS OF VARIANCE

                                SUM OF          MEAN
           SOURCE    DF        SQUARES        SQUARE       F VALUE       PROB>F

           MODEL      4     6.04854170    1.51213542         8.676       0.0001
           ERROR     28     4.88013127    0.17429040
           C TOTAL   32    10.92867297

               ROOT MSE       0.417481      R-SQUARE        0.5535
               DEP MEAN       3.105697      ADJ R-SQ        0.4897
               C.V.           13.44243

                                  PARAMETER ESTIMATES

                          PARAMETER        STANDARD      T FOR H0:
     VARIABLE  DF          ESTIMATE           ERROR    PARAMETER=0      PROB > |T|

     INTERCEP   1        7.74862694      0.86312958          8.977          0.0001
     UTL        1       -0.13996578      0.05620797         -2.490          0.0190
     SPA        1       -3.55383638      1.12221061         -3.167          0.0037
     ALF        1       -6.23622430      1.32043284         -4.723          0.0001
     TYPE       1        0.01672935      0.18130634          0.092          0.9271
```

The coefficient for the indicator variable TYPE is not statistically significant; therefore, there is no evidence of a difference in operating costs between the two types of airlines. The other coefficients are interpreted as before. In fact, they have values similar to the coefficients of the original model. This result is to be expected with the nonsignificant TYPE coefficient.

Although the inclusion of the indicator variable only estimates a difference in the average operating cost, this model does not address the possibility that the relationship of cost to the various operating factors differs between types. You can estimate this difference with a new model that includes variables that are products of the indicator and continuous variables.

A simple example can illustrate this principle. Assume the following model:

$$y = \beta_0 + \beta_1 x_1 + \beta_2 x_2 + \beta_3 x_1 x_2$$

where

x_1 is a continuous variable.

x_2 is an indicator variable with values 0 and 1 identifying two classes.

Then, for the first class (x_2=0), the model equation is as follows:

$$y = \beta_0 + \beta_1 x_1 \quad .$$

For the other class (x_2=1), the model equation is as follows:

$$y = (\beta_0 + \beta_2) + (\beta_1 + \beta_3)x_1 \quad .$$

In other words, the slope of the regression of y on x_1 is β_1 for the first class and $(\beta_1 + \beta_3)$ for the second class. Thus, β_3 is the difference in the regression coefficient between the two classes. This coefficient is sometimes referred to as a *shift* coefficient because it allows the coefficient to shift from one class to the next.

You can use this principle to estimate different regression coefficients for the two types of airlines for the three cost factors. First, create three additional variables in the DATA step:

```
UTLTP=UTL*TYPE;
SPATP=SPA*TYPE;
ALFTP=ALF*TYPE;
```

Then implement the regression:

```
PROC REG;
   MODEL CPM=UTL SPA ALF TYPE UTLTP SPATP ALFTP / VIF;
   ALLDIFF: TEST UTLTP,SPATP,ALFTP;
```

Three features of this implementation are of interest:

TYPE variable
: must be included. Leaving this variable out is equivalent to imposing the arbitrary requirement that the intercept is the same for models describing both types of airlines.

Variance inflation factors
: are not necessary for the analysis, but they do show an important feature of the results of this type of model.

TEST statement
: is used to test the null hypothesis that the three product coefficients are zero.

The results are shown in **Output 6.19**.

Output 6.19 Using an Indicator Variable for Slope Shifts

```
DEP VARIABLE: CPM
                              ANALYSIS OF VARIANCE

                              SUM OF         MEAN
         SOURCE     DF       SQUARES        SQUARE       F VALUE       PROB>F

         MODEL       7    8.35793768    1.19399110        11.611       0.0001
         ERROR      25    2.57073529    0.10282941
         C TOTAL    32   10.92867297

             ROOT MSE     0.3206703       R-SQUARE        0.7648
             DEP MEAN      3.105697       ADJ R-SQ        0.6989
             C.V.          10.32523

                              PARAMETER ESTIMATES

                        PARAMETER        STANDARD      T FOR H0:                          VARIANCE
   VARIABLE  DF          ESTIMATE           ERROR    PARAMETER=0      PROB > |T|          INFLATION

   INTERCEP   1       10.64436497      1.50431265          7.076          0.0001                  0
   UTL        1       -0.41589381      0.07547880         -5.510          0.0001         3.72250309
   SPA        1       -6.05483635      4.32444384         -1.400          0.1738        46.82543711
   ALF        1       -6.96499161      2.09708992         -3.321          0.0028         6.13091228
   TYPE       1       -4.07579888      1.72379750         -2.364          0.0261          232.92861
   UTLTP      1        0.41751264      0.09213361          4.532          0.0001        59.18560634
   SPATP      1        2.27294101      4.41724924          0.515          0.6114          110.06340
   ALFTP      1        0.61188332      2.41187768          0.254          0.8018        95.46334774

TEST: ALLDIFF     NUMERATOR:   0.769799  DF:    3   F VALUE:     7.4862
                  DENOMINATOR: 0.102829  DF:   25   PROB >F :    0.0010
```

The overall model statistics show that the residual mean square has decreased from 0.174 to 0.103 and the R-SQUARE value has increased from 0.5535 to 0.7648. In other words, allowing different coefficients for the two types of airlines has clearly improved the fit of the model. The test for the three additional coefficients given at the bottom of the output ($p=0.0010$) indicates that this improvement in the fit of the model is statistically significant.

The statistics for the individual coefficients show that among the added coefficients only UTLTP is statistically significant ($p<0.05$). In other words, the coefficient for the utilization factor is the only coefficient that can be shown to differ for the two types.

In checking the statistics for the other coefficients, note that neither SPA nor its change, or shift, coefficient (SPATP) is statistically significant ($p=0.1738$ and 0.6114, respectively). This appears to contradict the results for the previous models where the coefficient for SPA does indeed appear to be needed in the model. This apparent contradiction arises because it is not legitimate to make tests for lower order terms in the presence of higher order terms. This is true for this model just as it was in polynomial models (Chapter 5). In other words, if UTLTP is in the model, the test for UTL is not meaningful. One reason for this is the high degree of multicollinearity often found in models of this type. In this example the multicollinearity is made worse by a large difference in the sizes of planes (SPA) used by the two types of airlines. Therefore, the variance inflation factors involving TYPE and SPATP exceed 100.

For this reason, you can reestimate the model omitting the two nonsignificant shift terms, SPATP and ALFTP. The SAS statements are as follows:

```
PROC REG;
   MODEL CPM=UTL SPA ALF TYPE UTLTP;
   UTLTYPE1: TEST UTL+UTLTP;
```

You should note that the TYPE variable remains in the model since it is a lower order term in the model that still includes one product term. The TEST statement is included to ascertain the significance of the effect of UTL for the TYPE=1 airlines. The results are shown in **Output 6.20**.

Output 6.20 Final Equation for a Slope Shift

```
DEP VARIABLE: CPM
                                   ANALYSIS OF VARIANCE

                                  SUM OF          MEAN
              SOURCE    DF       SQUARES        SQUARE      F VALUE       PROB>F

              MODEL      5    8.33067049    1.66613410       17.315       0.0001
              ERROR     27    2.59800248    0.09622231
              C TOTAL   32   10.92867297

                 ROOT MSE     0.3101972      R-SQUARE       0.7623
                 DEP MEAN      3.105697      ADJ R-SQ       0.7183
                 C.V.          9.988007

                                   PARAMETER ESTIMATES

                            PARAMETER         STANDARD      T FOR H0:
         VARIABLE  DF        ESTIMATE            ERROR    PARAMETER=0      PROB > |T|

         INTERCEP   1     10.14142330       0.80789933         12.553          0.0001
         UTL        1     -0.42057223       0.07116286         -5.910          0.0001
         SPA        1     -3.87295233       0.83639693         -4.631          0.0001
         ALF        1     -6.40945771       0.98175412         -6.529          0.0001
         TYPE       1     -3.53349316       0.74133587         -4.766          0.0001
         UTLTP      1      0.42280941       0.08681854          4.870          0.0001

TEST: UTLTYPE1    NUMERATOR:     1.9E-04  DF:    1   F VALUE:     0.0019
                  DENOMINATOR: .0962223  DF:   27   PROB >F :    0.9653
```

The deletion of the two product variables has virtually no effect on the fit of the model. In this model all coefficients are highly significant, although the interpretation and statistics for the UTL coefficient may be clouded by the presence of its product variable.

This equation estimates that a one percent change in utilization decreases cost by 0.42 cents for the short-haul airlines. On the other hand, the corresponding effect for the long-haul airlines is the sum of the coefficients for UTL and UTLTP. This estimate is $-0.420572+0.422809=0.002237$, or almost zero. The TEST statement results show that utilization does not appear to be a cost factor for these lines.

The use of dummy variables is readily extended to more than one categorical variable. This use is also extended to situations where such variables have more than two categories. However, for all but the simplest cases (such as the one presented here), this type of analysis is more easily performed by PROC GLM, which automatically generates the dummy variables as well as the product variables. The use of PROC GLM is documented in *SAS System for Linear Models.*

Appendix
Statistical Review

A.1 INTRODUCTION

This appendix reviews some basic principles of statistical inference, linear models, and experimental design. The goal is to present concepts, terminology, and notation rather than a complete technical discussion. Regression principles and notation are discussed in detail in Chapter 1 of this book.

A.2 STATISTICAL INFERENCE

Inferential statistics is a formalized body of methods and techniques designed to gain information about a population using information on a sample of that population. It is an inductive process. The distribution of a population is characterized by *parameters*, which are unknown constants that are of interest. For example, the expected value of a random variable is some unknown population mean, a value that characterizes the center of the distribution in a certain sense. To estimate parameters, compute functions of the data such as sample means. A function of the data is termed a *statistic.* Statistics from a sample can be used to make inferences, or reasonable guesses, about the parameters of a population. Therefore, if you take a random sample of thirty students from a high school, the mean height for those thirty students is a reasonable estimate of the mean height of all the students in the high school. The methods of statistical inference in this book fall broadly into the categories of parameter estimation and hypothesis testing.

A.3 PARAMETER ESTIMATES

An *estimator* $\hat{\theta}$ of a parameter θ is a function of the data from a sample used to give estimates of the parameter. The *expected value* of $\hat{\theta}$ is the mean of the sampling distribution of $\hat{\theta}$ denoted either μ_{θ} or $E(\hat{\theta})$. If $E(\hat{\theta})=\theta$, then $\hat{\theta}$ is an *unbiased estimator* of θ. The *bias* in $\hat{\theta}$ is $E(\hat{\theta})-\theta$. The *variance* of $\hat{\theta}$, denoted either $\sigma_{\hat{\theta}}^2$ or $V(\hat{\theta})$, is the variance of the sampling distribution of $\hat{\theta}$. The *covariance* between two estimators, $\hat{\theta}_1$ and $\hat{\theta}_2$, is denoted $\text{cov}(\hat{\theta}_1,\hat{\theta}_2)$. The *standard error* of $\hat{\theta}$ is the standard deviation of the sampling distribution of $\hat{\theta}$ (that is, the square root of $\sigma_{\hat{\theta}}^2$).

When the standard error of $\hat{\theta}$ is estimated from the sample, it is usually referred to as the *standard error of the estimate* and is denoted $s_{\hat{\theta}}$. To illustrate, the sample mean $\bar{y}$ is an estimator of the population mean μ. Statistical theory states that the variance of $\bar{y}$ is σ^2/n where σ^2 is the population variance and n is the number of observations in the sample. Because the sample variance s^2 is an estimate of

σ^2, it follows that s^2/n is an estimate of σ^2/n. Thus, the estimated standard error of $\bar{y}$ is $\sqrt{s^2/n} = s/\sqrt{n}$.

The term *mean square* is used frequently to refer to a sample variance. Associated with mean squares are *degrees of freedom* (df), which can be thought of as units of information contained in the mean square.

Linear functions are an important class of functions in statistical inference. If $l_1, \ldots, l_k$ is a set of constants called *coefficients* and if $l = l_1y_1 + \ldots + l_ky_k$, then l is called a *linear function* or *linear combination* of $y_1, \ldots, y_k$. The quantities $y_1, \ldots, y_k$ may be parameters, observed data, or parameter estimates. Note that the sample mean $\bar{y}$ is a linear function of the observed values $y_1, \ldots, y_n$ with $l_1 = \ldots = l_n = n^{-1}$. However, the sample variance s^2 is not a linear function of $y_1, \ldots, y_n$. Linear combinations of parameters can be estimated by linear combinations of the respective estimators. If $\hat{\theta}_1, \ldots, \hat{\theta}_k$ are estimators of $\theta_1, \ldots, \theta_k$, then an estimator of $\theta = l_1\theta_1 + \ldots + l_k\theta_k$ is $\hat{\theta} = l_1\hat{\theta}_1 + \ldots + l_k\hat{\theta}_k$. The variance of $\hat{\theta}$ is $V(\hat{\theta}) = \Sigma_i l_i^2 V(\hat{\theta}_i) + \Sigma_{i \neq j} l_i l_j \text{cov}(\hat{\theta}_i, \hat{\theta}_j)$.

All of the inference procedures in this guide, including confidence intervals, prediction intervals, and tests of hypotheses, assume the data come from a normal distribution, although the issue is not raised in each discussion because modest violations of this assumption have little effect in the validity of the inferences.

You can test the hypothesis that the mean of the population is a given value by using the Student t statistic. The test statistic for the null hypothesis that the mean is some value μ_0 is constructed as follows:

$$t = (\bar{x} - \mu_0) / (s / \sqrt{n}) \quad .$$

This is the difference between the sample mean and the hypothesized mean divided by the standard error of the mean. This statistic has a Student t distribution under the null hypothesis and the assumption that the original population has a normal distribution. A t statistic usually has the form $t = (\hat{\theta} - \theta)/s_{\hat{\theta}}$ where $\hat{\theta}$ is a linear estimator of θ (that is, a linear function of the data) and the standard error $s_{\hat{\theta}}$ is based on a mean square whose probability of distribution is independent of $\hat{\theta}$. The degrees of freedom (df) for t are the same as for the mean square on which $s_{\hat{\theta}}$ is based. The degrees of freedom do not necessarily depend on the number of observations in the estimator $\hat{\theta}$.

As mentioned above, the t distribution may be a good approximation to the distribution of this statistic even when the underlying population is not normally distributed. Many SAS System procedures print out the t statistic for the hypothesis that the mean is zero (keyword T, labeled T FOR H0:MEAN=0) and a two-tailed significance probability or p value (keyword PRT, labeled PROB > |T|). The smaller the p value, the stronger the evidence against the null hypothesis.

The F statistic has the form $F = s_1^2/s_2^2$ where s_1^2 and s_2^2 are two independent mean squares. Df for the numerator and df for the denominator correspond to the df of s_1^2 and s_2^2, respectively.

The t and F probability distributions are used to construct confidence intervals for parameters and to conduct hypothesis tests about parameters. A $100(1-\alpha)$% confidence interval for a parameter θ, for which the estimator has a normal sampling distribution, has the following form:

$$\hat{\theta} \pm t_\alpha s_{\hat{\theta}} = (\hat{\theta} - t_\alpha s_{\hat{\theta}}, \hat{\theta} + t_\alpha s_{\hat{\theta}})$$

where $\hat{\theta}$ and $s_{\hat{\theta}}$ have their previous definitions and t_α is the two-tailed, α-level critical value for a t statistic with df equal to the df of $s_{\hat{\theta}}$. That is,

$$P[-t_\alpha < T < t_\alpha] = \alpha$$

where T is a t random variable.

A test of the hypothesis H_0: $\theta = a$, where a is specified, is usually based on the t statistic

$$t = (\hat{\theta} - a) / s_{\hat{\theta}} \quad .$$

If the true value of θ is far from a, then the estimator $\hat{\theta}$ will tend to be far from a. As a result, the absolute value of t will tend to be large; the larger the t value, the greater the incredibility associated with H_0. A measure of this incredibility is the probability that a t as large in absolute value as the observed t could have occurred if H_0 were true. This probability is called the *observed significance level*, or the *p value*, associated with the observed t.*

The F statistic, $F = s_1^2/s_2^2$, is used to test hypotheses of the form

$$H_0\text{: } \sigma_1^2 = \sigma_2^2$$

where σ_1^2 and σ_2^2 are the parameters estimated by s_1^2 and s_2^2. Typically, $E(s_1^2) = \sigma_1^2$ and $E(s_2^2) = \sigma_2^2$. In most uses of F, the only concern is to assess evidence that $\sigma_1^2 > \sigma_2^2$. The p value is computed as the probability that an F as large as the observed F could be obtained if $\sigma_1^2 = \sigma_2^2$. The folded form

$$F' = \max(s_1^2, s_2^2) / \min(s_1^2, s_2^2)$$

is used if the concern is to detect both $\sigma_1^2 > \sigma_2^2$ and $\sigma_1^2 < \sigma_2^2$.

The terms σ_1^2 and σ_2^2 do not necessarily stand for population variances. For example, consider

$$\sigma_1^2 = \sigma_2^2 + c_1 \sigma_A^2$$

where σ_2^2 and σ_A^2 are variances and c_1 is a known constant. In this application, testing H_0: $\sigma_1^2 = \sigma_2^2$ is really testing H_0: $\sigma_A^2 = 0$. In another example, consider

$$\sigma_1^2 = \sigma_2^2 + c_2 \Sigma_{i=1}^{k} \alpha_i^2$$

where the α_is are linear functions of means, σ_2^2 is a variance, and c_2 is a constant. In this application, H_0: $\sigma_1^2 = \sigma_2^2$ is equivalent to H_0: $\Sigma_{i=1}^{k} \alpha_i^2 = 0$, which in turn is equivalent to H_0: $\alpha_1 = 0, \ldots, \alpha_k = 0$.

The F distribution can also be used to construct confidence intervals, but this is not frequently done.

A.4 LINEAR MODELS

In this book the term *linear model* refers to an equation that relates a set of random variables, parameters, and deterministic variables and is linear in the parameters. Such an equation can usually be written in the form

$$y = \beta_0 + \beta_1 x_1 + \ldots + \beta_m x_m + \varepsilon$$

* Only a few years ago scientists used a set of tables to determine whether p was larger or smaller than a few specified values, typically p values of .01, .05, and .10. This reliance on tables was in part responsible for the accept/reject or significant/not significant approach to hypothesis testing that has pervaded the scientific literature. Now, it seems more desirable to report the exact p value rather than to declare "yea" or "nay" to indicate whether the test was statistically significant.

where

y is an observed random variable.

x_i are either deterministic or random variables whose values are known or observed.

β_i are either unknown constants or unobserved random variables.

ε is an unobserved random variable.

The y variable is called a *response* or *dependent variable*; the xs are called *independent variables.*

In statistics, linear models mathematically approximate the relation between y and the xs observed in a population. The variables y and $x_1, \ldots, x_m$ are each measured on a set of n sample entities. For instance, y is the height of a mature pine tree, and $x_1, \ldots, x_m$ is a set of variables, X, describing the growth of the tree, the species of the tree, and the conditions under which the tree was grown. The same variables are measured on n trees, and the data are recorded in an array:

$$\begin{bmatrix} y_1 & x_{11} & \cdots & x_{1m} \\ y_2 & x_{21} & \cdots & x_{2m} \\ \cdot & \cdot & & \cdot \\ \cdot & \cdot & & \cdot \\ \cdot & \cdot & & \cdot \\ y_n & x_{n1} & \cdots & x_{nm} \end{bmatrix} .$$

The subscript on y and the first subscript on x refer to the observation number; the second subscript on x identifies the x among $x_1, \ldots, x_m$.

The *method of least squares* is used to determine the model equation that best fits or best represents the data. This equation is obtained by finding the set of particular values of $\beta_0, \beta_1, \ldots, \beta_m$ that minimizes the sum of squares

$$\Sigma_{j=1}^{n}(y_j - (\beta_0 + \beta_1 x_{ij} + \ldots + \beta_m x_{mj}))^2 \quad .$$

These βs, denoted $\hat{\beta}_0, \hat{\beta}_1, \ldots, \hat{\beta}_m$, are called the *least-squares estimates* of $\beta_0, \beta_1, \ldots, \beta_m$.

The linear model is called a *regression model* when the x variables are quantitative variables like age and temperature. In some applications, each x variable measures a separate entity; in others, several x variables may be different functions of a common variable. For example, in *polynomial regression* $x_1=x$, $x_2=x^2, \ldots, x_m=x^m$ or successive powers of a variable x. Some regression variables can be controlled by an experimenter (for example, amount of applied nitrogen), and some are simply measured (for example, blood pressure).

Regression analysis attempts to construct a linear equation that is a good approximation (estimate) of the true (population) function relating the variables. This equation can be used to estimate the mean of a subpopulation corresponding to a particular set of xs, or it can be used to predict a future value to be drawn from a subpopulation corresponding to a particular set of xs. The parameter estimates, or functions of the parameter estimates, can also have a meaningful interpretation. For example, if y is daily electricity consumption and x_1 is the number of operations of the dishwasher per day, then β_1 is an estimate of electricity consumed per operation of the dishwasher.

Analyses associated with regression models using the SAS System comprise the topics covered in this book.

Linear models also apply to situations where independent variables are categorical and not quantitative, as they are in regression analysis. This type of model is referred to as an *analysis-of-variance*, or ANOVA, model. In analysis of variance, the independent variables are called *factors*, or *treatments*. Different values of the variable are referred to as *levels* of the factor and define the *treatment group* to which each observation belongs. The means and variances of the y variable observed in each treatment group are the focus of the analysis.

A one-factor ANOVA model can be written

$$y_{ij} = \mu + \alpha_i + \varepsilon_{ij}$$

where

y_{ij} is the observed value of a dependent variable of interest.

μ is the grand mean.

α_i represents the difference between the observed value of y and the grand mean due to the effect of one independent variable.

ε_{ij} is experimental error.

When the levels of a factor represent all levels of interest to the investigator, the model is a *fixed-effects* ANOVA model. When the levels of a factor represent a sample of individual units from a population of levels (for example, classes in a school, operators of a machine, replications of an experiment), the model is a *random-effects* ANOVA model. The random-effects or *variance-component* model is a linear model in which some of the α_i are defined to be random variables. In this application the α_i are not estimated; instead, their variances are estimated.

A.5 EXPERIMENTAL DESIGN

A researcher uses principles of experimental design to plan experiments that provide the most information with available resources. This usually involves two basic processes: *treatment selection* and *error control*.

Treatment selection is the choice of experimental variables, treatments, levels, and settings of these variables to include in the experiment. Changes in the response variable due to changes in the treatment level are called *effects* of the factor or treatment.

For example, an experiment to investigate the effects of different irrigation methods (sprinkle, spray, and drip) on ornamental nursery plants has METHOD OF IRRIGATION as an independent variable or factor. Individual methods (sprinkle, spray, and drip) are the levels of this factor.

Consider expanding the experiment to examine the effects of different amounts of water, say 10, 20, 30, and 40 liters per day. AMOUNT OF WATER becomes a second factor; levels of this factor are the different amounts of water. The two factors can be referred to as METHOD and AMOUNT. The AMOUNT factor is *continuous*, or *quantitative*, because its levels (10, 20, 30, 40) are quantitative values selected from a continuum of possible values. The METHOD factor is *discrete*, or *qualitative*, because its levels (sprinkle, spray, and drip) do not represent quantities.

Combinations of METHOD and AMOUNT make up the treatments in the experiment: 10 liters of water applied using sprinkle is one treatment, and 20 liters of water applied using drip is another. An experiment containing all possible combinations of levels of factors is called a *factorial experiment*; other selections of the treatments depend on the objectives of the experiment. A group of plants that will not be watered in the experiment can be included; such a zero rate of a continuous variable is called a *control.*

Conventional experiments assign treatments to experimental units according to some random method. Treatments are assigned in a way that will reduce error variation from contributing to the standard error of the estimates of the parameters. One method of error control is called *blocking*; experimental units are grouped into *blocks* so that units in the same block are as homogeneous as possible. In the most common type of blocking, each treatment is randomly assigned to a unit in each block; the result is that identifiable variation between blocks is removed from differences between treatment means.

Suppose you want to estimate the difference in mean weight gains of swine on two DIETS (DIET A and DIET B). In the pen are pigs from several LITTERS, with two pigs from each litter. You suspect that weight gains of litter mates tend to be more nearly the same than weight gains of pigs from different litters; that is, different litters are a likely source of identifiable variation, so you block the pigs according to LITTER. You feed DIET A to a randomly selected pig in each litter and DIET B to the other pig. A linear model for the weight gains may take the form

$$y_{ij} = \mu_i + \delta_j + \varepsilon_{ij}$$

where

y_{ij} is the weight gain of the pig in LITTER j that receives DIET i.

μ_i is the population mean of weight gains of pigs receiving DIET i.

δ_j is the random quantity associated with all pigs of LITTER j, the variation between LITTERS.

ε_{ij} is the random quantity associated with the particular pig in LITTER j that received DIET i, the variation within LITTERS.

The random quantities δ_j and ε_{ij} are usually taken to be independent random variables with respective means $\mu_\delta=\mu_\varepsilon=0$ and respective variances σ_δ^2 and σ_ε^2. Then the sample mean weight gain of pigs receiving DIET i is

$$\bar{y}_{i.} = \mu_i + \bar{\delta}. + \varepsilon_{i.} \quad .$$

The difference between the means of the pigs receiving the two diets $\bar{y}_{A.}-\bar{y}_{B.}$ can be written

$$\begin{aligned} \bar{y}_{A.} - \bar{y}_{B.} &= (\mu_A + \bar{\delta}. + \bar{\varepsilon}_{A.}) - (\mu_B + \bar{\delta}. + \bar{\varepsilon}_{B.}) \\ &= (\mu_A - \mu_B) + (\bar{\varepsilon}_{A.} - \bar{\varepsilon}_{B.}) \end{aligned}$$

where a dot (.) in place of a subscript indicates summation over the subscript.

Note that the difference does not contain any δs. The two pigs from each litter have the same δ_j value so that the average δ value for DIET A pigs is the same as the average δ value for DIET B pigs. Because the average δ value, denoted $\bar{\delta}.$, is subtracted out of the right side of the equation, the standard error of the estimator, $\bar{y}_{A.}-\bar{y}_{B.}$, of $\mu_A-\mu_B$ does not contain a term involving σ_δ^2. In fact, the standard error is

$$\sqrt{2\sigma_\varepsilon^2 / n}$$

where n is the number of litters. By identifying potential variation due to δ and the differences between litters and by making a judicious treatment assignment, you have eliminated between-litter variation from the estimator.

References

Allen, D.M. (1970), "Mean Square Error of Prediction as a Criterion for Selecting Variables," *Technometrics*, 13, 469-475.

Belsley, D.A., Kuh, E., and Welsch, R.E. (1980), *Regression Diagnostics*, New York: John Wiley & Sons, Inc.

Dickens, J.W. and Mason, D.D. (1962), "A Peanut Sheller for Grading Samples: An Application in Statistical Design," *Transactions of the ASAE*, Volume 5, Number 1, 42-45.

Draper, N.R. and Smith, H. (1981), *Applied Regression Analysis*, Second Edition, New York: John Wiley & Sons, Inc.

Draper, N.R. and Van Nostrand, R. C. (1979), "Ridge Regression and James-Stein Estimation: Review and Comments," *Technometrics*, 21, 451-466.

Freund, R.J. and Minton, P.D. (1979), *Regression Methods*, New York: Marcel Dekker, Inc.

Freund, R.J. and Littell, R. (1986), *SAS System for Linear Models, 1986 Edition*, Cary, NC: SAS Institute Inc.

Fuller, W.A. (1976), *Introduction to Statistical Time Series*, New York: John Wiley & Sons, Inc.

Fuller, W.A. (1986), "Using PROC NLIN for Time Series Prediction," *Proceedings of the Eleventh Annual SAS Users Group International Conference*, Cary, NC: SAS Institute Inc., 63-68.

Graybill, F. (1976), *Theory and Application of the Linear Model*, Boston: PWS and Kent Publishing Company, Inc.

Johnson, R.A. and Wichern, D.W. (1982), *Applied Multivariate Statistical Analysis*, Englewood Cliffs, NJ: Prentice Hall.

Kvalseth, T.O. (1985), "Cautionary Note about R^2," *The American Statistician*, 39, 279-286.

Mallows, C.P. (1973), "Some Comments on C(p)," *Technometrics*, 15, 661-675.

Montgomery, D.C. and Peck, E.A. (1982), *Introduction to Linear Regression Analysis*, New York: John Wiley & Sons, Inc.

Morrison, D.F. (1976), *Multivariate Statistical Methods*, Second Edition, New York: McGraw-Hill Book Co.

Myers, R.H. (1976), *Response Surface Methodology*, Blacksburg, Virginia: Virginia Polytechnic Institute and State University.

Myers, R.H. (1986), *Classical and Modern Regression with Applications*, Boston: PWS and Kent Publishing Company, Inc.

Rawlings, J.O. (1988), *Applied Regression Analysis: A Research Tool*, Pacific Grove, California: Wadsworth & Brooks/Cole Advanced Books & Software.

Sall, J.P. (1981), "SAS Regression Applications," Revised Edition, SAS Technical Report A-102, Cary, NC: SAS Institute Inc.

SAS Institute Inc. (1984), *SAS/ETS User's Guide, Version 5 Edition*, Cary, NC: SAS Institute Inc.

SAS Institute Inc. (1985), *SAS/IML User's Guide, Version 5 Edition*, Cary, NC: SAS Institute Inc.

SAS Institute Inc. (1985), *SAS/STAT Guide for Personal Computers, Version 6 Edition*, Cary, NC: SAS Institute Inc.

SAS Institute Inc. (1985), *SAS User's Guide: Basics, Version 5 Edition*, Cary, NC: SAS Institute Inc.

SAS Institute Inc. (1985), *SAS User's Guide: Statistics, Version 5 Edition*, Cary, NC: SAS Institute Inc.

SAS Institute Inc. (1985), *SAS Introductory Guide for Personal Computers, Version 6 Edition*, Cary, NC: SAS Institute Inc.

Searle, S.R. (1971), *Linear Models*, New York: John Wiley & Sons, Inc.

Smith, P.L. (1979), "Splines as a Useful and Convenient Statistical Tool," *The American Statistician*, 33, 57-62.

Steel, R.G.B. and Torie, J. H. (1980), *Principles and Procedures of Statistics*, Second Edition, New York: McGraw-Hill Book Co.

Index

P

Q

R

S

Proofreading and text entry support are performed in the **Technical Writing Department** by **Amy E. Ball**, **Kimberly I. Barber**, **Gina A. Eatmon**, **Rebecca A. Fritz**, **Lisa K. Hunt**, **Caroline T. Powell**, **Beth L. Puryear**, **Drew T. Saunders**, **W. Robert Scott**, and **Harriet J. Watts**, under the supervision of **David D. Baggett**. **Gigi Hassan** is index editor.

Production is performed in the **Graphic Arts Department**. Composition was provided by **Blanche W. Phillips**. Text composition programming was provided by **Craig R. Sampson**.

Creative Services artist **Lisa Clements** provided illustrations under the direction of **Jennifer A. Davis**.

Your Turn

If you have comments about SAS software or the *SAS System for Regression, 1986 Edition*, please send us your ideas on a photocopy of this page. If you include your name and address, we will reply to you.

Please return to the Publications Division, SAS Institute Inc., SAS Circle, Box 8000, Cary, NC 27512-8000.